Nucleic Acid Biosynthesis

METHODS IN MOLECULAR BIOLOGY

Edited by

ALLEN I. LASKIN
*ESSO Research and Engineering
Company
Linden, New Jersey*

JEROLD A. LAST
*Harvard University
Cambridge, Massachusetts*

ADDITIONAL VOLUMES IN PREPARATION

Nucleic Acid Biosynthesis

EDITED BY

Allen I. Laskin and Jerold A. Last

ESSO RESEARCH AND
ENGINEERING COMPANY
LINDEN, NEW JERSEY

HARVARD UNIVERSITY
CAMBRIDGE, MASSACHUSETTS

MARCEL DEKKER, INC. New York 1973

MARCEL DEKKER, INC.

95 Madison Avenue, New York, New York   10016

LIBRARY OF CONGRESS CATALOG CARD NUMBER: 72-89529

ISBN: 0-8247-6008-5

Printed in the United States of America

PREFACE

Every laboratory research worker has experienced the many frustrations associated with attempting to duplicate a procedure described in the conventional scientific literature. Rarely does a paper describe the experimental methods in a manner that allows one readily to reproduce them in his own laboratory. This problem has resulted in the appearance of a variety of books, manuals, etc., on "methods," "techniques," and "procedures." In many instances, however, the descriptions found in such books may be no easier to follow than those in the original literature; they may be "buried" in a volume of unrelated techniques, and often may be in a book with a rather high cost.

Methods in Molecular Biology represents an attempt to provide small, relatively inexpensive, topically organized volumes, which might be particularly beneficial to new workers in a field, to graduate students beginning a problem, to new technicians, etc.

The authors were asked to write descriptions of the methods used in a particular area as critically as possible, and whenever appropriate, to discuss such things as: why a

particular approach was taken, <u>why</u> a particular reagent was
used, what alternatives are feasible and acceptable, and what
to do "if things go wrong."

In treating the subject of nucleic acid biosynthesis, we
deliberately emphasized RNA, at the expense of DNA, due to the
current uncertainties and the rapid obsolescence of information
in the latter field. A future volume of this series will deal
with DNA biosynthesis in detail. In this book, the preparation
and assay of 5 S RNA and single-stranded RNA from animal
viruses, of various RNA polymerases, RNA-directed DNA polymerases, DNA polymerase II, RNase H, and of substrates for
polymerase assays are emphasized. The preparation and assay of
bacteriophage RNA, ribosomal RNA, transfer RNA, and mammalian
messenger RNA (for hemoglobin) have already been detailed in
Volumes 1 and 2 of this series. Three additional chapters are
included: on ways to dissociate inhibition of RNA and protein
synthesis <u>in vivo</u> (to allow study of the one process without
interference from the other), on preparation of extracts from
Krebs ascites tumor cells (which promises to be a universal
system for assay by translation of mRNAs from eukaryotes and
viruses), and on methods to synchronize mammalian cells in
tissue culture to allow study of DNA synthesis <u>in vivo</u>.

CONTRIBUTORS TO THIS VOLUME

H. AVIV, Laboratory of Molecular Genetics, National Institute
of Child Health and Human Development, National Institutes
of Health, Bethesda, Maryland

IRA BERKOWER, Department of Developmental Biology and Cancer,
Albert Einstein College of Medicine, Bronx, New York

DAVID H. L. BISHOP, Institute of Microbiology, Rutgers University, New Brunswick, New Jersey

I. BOIME, Laboratory of Molecular Genetics, National Institute
of Child Health and Human Development, National Institutes
of Health, Bethesda, Maryland

HERBERT L. ENNIS, Roche Institute of Molecular Biology, Nutley,
New Jersey

PETER HERRLICH, Max-Planck-Institut für Molekulare Genetik,
Berlin-Dahlem, West Germany

JEROLD A. LAST, Harvard University, Cambridge, Massachusetts

RUSTY J. MANS, Department of Biochemistry, University of
Florida, Gainesville, Florida

CHARLES A. PASTERNAK, Department of Biochemistry, Oxford
University, Oxford, England

ROBERT RÖSCHENTHALER, Institut für Hygiene und Medizinische,

   Mikrobiologie der Technischen Hochschule München, Munich,

   West Germany

T. SREEVALSAN, Department of Microbiology, Georgetown Univer-

   sity Schools of Medicine and Dentistry, Washington, D.C.

REED B. WICKNER, Department of Developmental Biology and

   Cancer, Albert Einstein College of Medicine, Bronx,

   New York

THEODORE P. ZACHARIA, National Academy of Sciences, Washington,

   D.C.

CONTENTS

Chapter 3.   RNA POLYMERASE IN HIGHER PLANTS

Rusty J. Mans

Chapter 4.   RNA-DEPENDENT DNA POLYMERASES

Jerold A. Last and Theodore P. Zacharia

Chapter 5.   RIBONUCLEASE H OF ESCHERICHIA COLI

Ira Berkower

Chapter 6. DNA POLYMERASE II

Reed B. Wickner

Chapter 7. DISSOCIATION OF PROTEIN AND RIBONUCLEIC
            ACID SYNTHESIS IN VIVO

Herbert L. Ennis

Chapter 8. CELL-FREE SYNTHESIS OF VIRAL PROTEINS IN
            THE KREBS II ASCITES TUMOR SYSTEM

I. Boime and H. Aviv

Chapter 9.   PREPARATION OF 5 S RNA

Robert Röschenthaler and Peter Herrlich

Chapter 10.   SYNCHRONIZATION OF MAMMALIAN CELLS
              BY SIZE SEPARATION

Charles A. Pasternak

Chapter 1

RNA-DEPENDENT RNA POLYMERASES

David H. L. Bishop

Institute of Microbiology
Rutgers University
New Brunswick, New Jersey

I.   INTRODUCTION

The process of RNA-templated RNA synthesis involves several well-defined steps: recognition of the template by the enzyme, initiation of RNA synthesis, incorporation of precursor nucleotides into RNA, and removal of the product strand from the transcription complex. How RNA-templated synthesis is studied _in vitro_ will be described. Also, procedures involving product analysis by a variety of techniques, including nearest neighbor analysis, sequence studies, hybridization, gel electrophoresis, etc., will be discussed. Since radioisotope-labeled ribonucleotide triphosphates usually are employed to monitor product RNA initiation and synthesis, procedures for their syntheses and purification will also be described.

Several systems involving RNA transcription of RNA virus genomes have been described; each of these will be discussed briefly concerning what is known about the transcription process, how the information was derived, and the function of the transcribed product.

Most of our information on RNA transcription from RNA templates has been obtained with the RNA bacteriophage Qβ replicase, and a procedure for its purification will be described. In the last few years, evidence for RNA transcription of animal RNA virus genomes (reovirus, vesicular stomatitis (VSV), influenza, Newcastle disease, and Sendai viruses) has been obtained and it has become apparent that RNA transcription is a major initial process in the cell-infection process by many RNA viruses. Since we do not yet know of any nonviral RNA-dependent RNA transcription process in uninfected animal, plant or bacterial cells, these viral RNA-transcription processes appear to be unique, and are therefore suitable targets for study and possibly for eventual selective antiviral chemotherapy.

II. SYNTHESIS OF LABELED RIBONUCLEOSIDE TRIPHOSPHATES

A. Preparation of [α -^{32}P]Ribonucleoside Triphosphates

The method used to prepare [α-^{32}P]-labeled ribonucleoside triphosphates is based on that described by Symons [1] with modifications to obtain triphosphates in high yields that reproducibly exhibit high specific activity as well [2, 3, 4].

The technique is described in complete detail to allow interested investigators to prepare their own labeled triphosphates without incurring the prohibitive cost of purchasing similar, commercially synthesized products.

1. Purification of[^{32}P]Phosphoric Acid. Carrier-free [^{32}P]phosphoric acid (150 mCi) in 0.01 N HCl (ICN, Irvine, Calif.) is mixed with 0.1 ml of 0.1 M H_3PO_4 together with four drops (about 0.2 ml) of 5 N NH_4OH (to make the solution alkaline). The ammonium phosphate is loaded on a 5-ml column of Dowex-1 (Cl$^-$; X8, 200 to 400 mesh, BioRad, Richmond, Calif.) and is washed with 10 ml of distilled water. After absorption of the phosphate, the column is eluted with 0.05 M LiCl in 0.01 N HCl, and 3-ml fractions are collected. The fractions are monitored for content of ^{32}P using a hand geiger counter. To each fraction, 2 drops (about 0.1 ml) of a saturated solution of $BaCl_2$ are added together with 0.3 ml of 5 N NH_4OH. In the fractions that contain most of the radioactivity, a bluish-white flocculent precipitate is observed. Subsequent fractions often give a heavy white precipitate upon addition of the $BaCl_2$, which is possibly due to silicates or sulfates present in the starting materials. The initial fractions are pooled and mixed with two volumes of methanol in a Corex 30-ml centrifuge tube (Corning, Corning, N. Y.). The barium phosphate precipitate is recovered by centrifugation at 1,000 X g for 10 min, and the supernatant fluids are discarded. Barium is exchanged for hydrogen by suspending the pellet in 2 to 3 ml of water and shaking with 1 ml

of Dowex 50 (H$^+$) beads (Dowex-50W X8, 20 to 50 mesh, BioRad).

After the pellet has dissolved, the supernatant fluid, contain-

ing the phosphoric acid, is carefully transferred to a 50-ml

round-bottom flask, and the beads are re-extracted with an addi-

tional 2 ml of water. The combined fluids are evaporated to

dryness under vacuum at 37°C by using a Calab flash evaporator

(Calab, Oakland, Calif.) and a dry ice-methanol trap for col-

lecting the effluent water vapors. After drying, 4 ml of water

are added through the Calab solvent addition insert, and the

solution again is evaporated to dryness. This process is re-

peated to ensure complete removal of HCl from the phosphoric

acid.

  2.   Synthesis of [^{32}P] Ribonucleoside 5' Monophosphates.

Triethylamine, 0.1 ml (redistilled and stored, anhydrous, over

calcium hydride) is added together with approximately 3 ml of

acetonitrile (redistilled and stored, anhydrous, over molecular

sieves) to the [^{32}P]phosphoric acid. The mixture is evaporated

to dryness. An additional 3 ml of anhydrous acetonitrile is

added and removed by evaporation under vacuum. Approximately

150 mg of the protected isopropylidene derivative of the desired

ribonucleoside (P.-L. Biochemicals Inc., Milwaukee, Wisconsin)

is dissolved (or suspended) in a minimum volume of anhydrous

acetonitrile (3 to 8 ml) and then added to the flask and evapo-

rated to dryness. The flask is removed from the evaporator and

sealed with a silicone rubber stopper. The following successive additions are made (with thorough mixing at each stage): 3 ml

of anhydrous dimethylsulfoxide (Me_2SO, distilled and stored,

anhydrous, over molecular sieves), 4 drops (about 0.2 ml) of

anhydrous triethylamine and 4 drops (about 0.2 ml) of anhydrous

trichloroacetonitrile (distilled and stored, anhydrous, over

molecular sieves). The stoppered flask is then incubated at

37°C for 30 min to allow synthesis of the 5' monophosphate

derivative of the ribonucleoside. The contents usually turn

brown; if not, further triethylamine is added to neutralize con-

taminating HCl, and the mixture is reincubated. The contents

are transferred to a 30-ml Corex centrifuge tube, using 1 ml of

Me_2SO to rinse the flask, followed by 2 ml of water. Phosphoric

acid (0.5 ml of a 0.1 M solution) is added (as a coprecipitant)

followed by 10 drops (about 0.5 ml) of 0.1 M LiOH (to make the

solution alkaline). Four drops (about 0.2 ml) of saturated

$BaCl_2$ are added to precipitate all the phosphates, followed by

3 to 4 volumes of ethanol. The mixture is allowed to stand at

4°C for 30 min, and then centrifuged at 1,000 X g for 20 min.

The supernatant is carefully checked for content of radioactivity.

(If more than 5% of the label remained in the supernatant, it

is recentrifuged at a greater speed to recover unsedimented

radioactive material.) Subsequently, the supernatant fluid is

discarded, and the pellet is suspended in 5 ml of ethanol and

recentrifuged to remove residual Me_2SO. The alcoholic super-

natant fluid is discarded, and the barium phosphates are sus-

pended in 2 ml of water and treated with Dowex-50 (H^+) beads (as described above) to exchange the barium for hydrogen. After washing the beads with an additional 2-ml volume of water, the supernatant fluids are transferred to a 50-ml round-bottom flask, adjusted to 5 N acetic acid using glacial acetic acid, and heated at 100°C in a steam bath for 45 min to remove the protecting isopropylidene group. After cooling, the acetic acid is removed using the Calab flash evaporator, and the residues are suspended in 5 ml of water in preparation for enzymatic conversion of the 5' ribonucleoside monophosphate to its corresponding triphosphate using nucleoside phosphate kinases and an energy generating system.

   3. Preparation of Ribonucleoside Mono- and Diphosphate Kinases. A variation of the procedures of Bresler [5] and Hurlbert and Furlong [6] has been used to isolate a protein fraction from Escherichia coli which contains both nucleoside mono- and diphosphate kinase activities.

   Streptomycin sulfate supernatant fluids (the fraction 11 supernatant fluid described by Lehman et al. [7]), are prepared and stored at -20°C. A portion of these streptomycin superna- tant fluids (1,500 ml) is thawed, precipitated material is re- moved by centrifugation for 20 min at 20,000 X g, and 0.21 g of $(NH_4)_2SO_4$ is added per ml. After 10 min at 4°C, the precipitate is removed by centrifugation (10 min at 20,000 X g). An addition- al 0.09 g of $(NH_4)_2SO_4$ per ml of supernatant fluid is added, the mixture allowed to stand at 4°C for 10 min, and the precipitate removed by centrifugation as before. A third quantity of $(NH_4)_2SO_4$ (0.05 g per ml) is added, and, after 10 min at $4°C$, the precipitate is collected by centrifugation. This precipitate is dissolved in 0.05 M glycine buffer, 0.2 mM ethylenediaminetetraacetic acid, 2 mM glutamine (pH 8.0) to give a protein concentration of 11 mg per ml, as determined from the absorbance, $\underline{A}_{260}$ and $\underline{A}_{280}$. The solution is mixed with 0.15 ml of alumina Cγ gel (5% solids, Sigma Chemical Co., St. Louis, Mo.) per ml and gently stirred at $4°C$ for 10 min. The gel is removed by centrifugation at 20,000 X g for 10 min, and the supernatant is made 95% with respect to $(NH_4)_2SO_4$. After 10 min at $4°C$, the precipitate is collected by centrifugation as described above and dissolved in the glycine buffer to give a protein concentration of about 3 mg per ml. One-ml portions are stored at $-70°C$. In the conversion of adenosine 5' monophosphate and cytidine 5' monophosphate to their corresponding triphosphates, the enzyme preparation is dialyzed for 4 hr against 500 ml of 0.4 M NH_4Cl, 0.01 M Tris·HCl buffer (pH 8.0) prior to use in order to remove residual traces of ammonium sulfate. Otherwise, the sulfate elutes with these triphosphates from the Dowex-1 (Cl^-) column and gives an unmanageable precipitate on addition of the $BaCl_2$ (see below). With the other 5' monophosphates the enzyme preparation is used directly.

4. Conversion of Ribonucleoside Monophosphates to Corresponding Triphosphates. The reaction mixture used for converting the $[^{32}P]$ribonucleoside 5'monophosphates to triphosphates

contains (in 10 ml final volume):   500 µmoles of Tris·HCl buffer (pH 7.4), 10 µmoles of ATP, 40 µmoles of phosphoenopyruvate (sodium salt), 100 µmoles of $MgCl_2$, 1.2 mg of pyruvate kinase (Sigma Chemical Co.), 3 mg of the nucleoside phosphate kinase protein and the ^{32}P-labeled ribonucleoside 5'monophosphate. When [^{32}P]AMP is converted, only 1 µmole of unlabeled ATP is added to the reaction mixture. The reaction is incubated at 37°C for 40 min and then diluted with an equal volume of water.

5. Purification of the [α-^{32}P]Ribonucleoside Triphosphate. The diluted reaction mixture containing the [α-^{32}P]-ribonucleoside triphosphate is loaded on a 10-ml column of Dowex-1 beads (Cl$^-$; X8, 200 to 400 mesh). After the column is washed with 20 ml of water, followed by 30 ml of 0.05 M LiCl in 0.01 N HCl to remove proteins and [^{32}P]phosphoric acid, the triphosphates are eluted with 360 ml of a linear salt gradient in 0.01 N HCl. The gradients used are: 0.05 to 0.20 M LiCl for CTP, or to 0.35 M LiCl for ATP and GTP, or to 0.42 M LiCl for UTP. Fractions (8 ml) are monitored for radioactivity and their $\underline{A}_{260}$ and $\underline{A}_{280}$ are determined. The fractions containing labeled triphosphate are pooled, neutralized with 1 M LiOH, and concentrated by evaporation on the Calab evaporator to a volume of approximately 5 ml. After evaporation, the solution is transferred to a 30-ml Corex centrifuge tube. Five drops (about 0.25 ml) of saturated $BaCl_2$ and 4-5 volumes of ethanol are added. Insoluble barium triphosphate is collected by centrifugation at

1,000 X g, and barium is exchanged for hydrogen with Dowex-50 (H^+) beads as described previously. After the beads are washed, the combined supernatant fluids are neutralized with LiOH and evaporated to dryness. Finally the [α-^{32}P]ribonucleoside triphosphate is dissolved in water to give a concentration of approximately 1 μmole per ml (as determined from the A_{260}), and the specific activity is determined on a suitable dilution. All operations are performed behind 0.5-in. thick Plexiglass shields to protect against radiation hazards. The purity of the triphosphates is determined by paper electrophoresis at pH 3.5 (see later). The yield for all triphosphates is usually around 20% with respect to the initial ^{32}P label used, and the specific activity ranges between 5 and 10 mCi per μmole.

6. Comments. By monitoring the synthesis at various stages, we have usually found that the enzymatic conversion of any of the monophosphates to corresponding triphosphates is between 90 and 95% complete; also, the loss of label from monophosphate to free phosphate is negligible at the acid-hydrolysis stage, and the conversion of [^{32}P]phosphate to nucleoside monophosphate in the Me$_2$SO solution usually is around 90 to 95%. Only 20% of the initial label appears in the final product, presumably because solubilization of the phosphoric acid into the Me$_2$SO phase is only partial and that most of the phosphate remains attached to the glass during the Me$_2$SO incubation. This interpretation is somewhat validated by

the fact that almost all of the label in the Me_2SO solution is
in the nucleoside monophosphate, but that this amount only
accounts for 20 to 30% of the initial label, whereas after
washing out the flask, 50 to 80% of the total label is in the
form of free phosphate.

B. Preparation of $[\beta,\gamma-^{32}P]$Ribonucleoside Triphosphates

The synthesis of β, γ-labeled triphosphates involves con-
densation of $[^{32}P]$pyrophosphate with the corresponding nucleotide
morpholydate, and purification of the triphosphate from these
precursors [3]. The procedure described below is based on one
described by Wehrli et al. [8].

1. Synthesis of $[\beta,\gamma-^{32}P]$Ribonucleoside Triphosphates.

One hundred-200 mCi of $[^{32}P]$pyrophosphate, pyridine salt (ICN,
Irvine, California), is mixed with 1 ml of pyridine and 0.03 ml
of tributylamine. The mixture is evaporated to dryness under
reduced pressure and further dried five times by addition and
subsequent evaporation of 3 ml of redistilled, anhydrous pyridine
using the Calab evaporator described previously. The pyridine is
then removed by adding and evaporating 5 ml of anhydrous benzene.
One ml of dry, redistilled Me_2SO is then added and the flask
sealed. In a separate flask 100 mg of the desired ribonucleoside
5' monophosphate morpholydate (Sigma, St. Louis, Mo.) is dissolved or suspended in 3 ml of anhydrous pyridine and the pyridine
is removed under reduced pressure as before. This drying procedure is repeated five times and all traces of pyridine are

removed by adding and subsequently evaporating 3 ml of dry

benzene. To this dried substrate is quickly added the solution

of [^{32}P]pyrophosphate and, after dissolving, the mixture is

left for two days at 30°C in a sealed flask. The maintenance

of an anhydrous condition is critical to the success of the con-

densation process.

2. Purification of the[β, γ-^{32}P]Ribonucleoside Triphos-

phate. The labeled triphosphate is purified from the reaction

products by chromatography on Dowex-1-Cl⁻ as described for the

[α-^{32}P]ribonucleoside triphosphates. Since polyphosphates elute

from the Dowex column with, and after, UTP, and pyrophosphate

elutes between CTP and ATP, a further purification from these

polyphosphates is also necessary. One procedure that can be

used successfully is that of paper electrophoresis in 0.1 M

ammonium acetate, pH 3.8. Electrophoresis is conducted on

Whatman 54 paper, at 5000 V and 50 mA for 90 min. The triphos-

phates move much more slowly than pyrophosphate--and can be

identified by radioautography or their ultraviolet absorption.

The triphosphate is then eluted from the paper with water, the

paper floc is removed by centrifugation, and the triphosphate

is purified by barium precipitation and is finally converted to

the lithium or sodium salt as described above.

3. Comments. Usually a yield of between 20 and 50% is

obatined, depending on the batch of [^{32}P]pyridine pyrophosphate.

The final specific activity of the product is between 10 and 20

mCi per μmole.

C. Preparation of [γ-^{32}P]Ribonucleoside Triphosphates

The synthesis of γ-^{32}P-labeled ribonucleoside triphosphates can be accomplished enzymatically [4], and the procedure
described below is based on the methodology used by Glynn and
Chapell [9].

1. Synthesis of [γ-^{32}P]Ribonucleoside Triphosphates.

The reaction mixture (10 ml) contains 2 ml of 1 M Tris·HCl
buffer (pH 7.4), 0.03 ml of 1 M MgCl$_2$, 0.03 ml of 1 M 2-mercapto-
ethanol, 0.06 ml of 0.01 M phosphoric acid, 0.12 ml of 0.05 M
nucleoside triphosphate, 5 ml of H$_3$^{32}PO$_4$ (carrier-free, 40 mCi,
I.C.N., Irvine, California) - neutralized to pH 7 with NaOH),
1.00 ml of 0.1 M sodium 3-phosphoglycerate, 0.03 ml of 5 mM NAD,
0.06 ml of 3-phosphoglycerate kinase (Sigma, 10 mg/ml), 0.10 ml
of glyceraldehyde phosphate dehydrogenase (Sigma, 10 mg/ml) and
distilled water to give a final volume of 10 ml. For the synthesis of labeled ATP and CTP, the enzymes are dialyzed before
use for 4 hr against 0.4 M NH$_4$Cl, 0.01 M Tris·HCl buffer (pH 7)
to remove the (NH$_4$)$_2$SO$_4$ (see above).

The mixture is incubated at 37°C for 30 min, diluted to
20 ml with water, and loaded on a Dowex-1-Cl$^-$ column as described.
The elution of the column, purification of the triphosphates
through barium precipitation, and recovery of the triphosphates
as their lithium or sodium salts is described above.

2. Comments.

The yield is around 80 to 90% with respect
to the ^{32}P conversion, and gives a triphosphate preparation having a specific activity of 10-30 mCi per μmole.

D. Quality Control and Monitoring the Integrity of Tri-
phosphates on Storage

The synthesis and radiochemical purity of triphosphates--
both during and after purification, as well as in reaction mix-
tures--can be quickly and conveniently assayed by high-voltage
paper electrophoresis. We separate mono-, di-, tri-, and tetraphosphates of a nucleoside from free phosphate and pyrophosphate
with sodium citrate buffer (pH 3.5). With 0.05 M sodium citrate
(pH 3.5) and Whatman 54 paper, separation can be easily attained
in 60 min of electrophoresis at 40 mA and 6000 V. Nucleosides
and nucleotides, when present in sufficient quantities, can be
visualized through their absorption of ultraviolet light. Additionally, the presence of radiochemically labeled compounds can
be determined by radioautography with Kodak no-screen X-ray
film, and the amounts of label in any spot may be determined by
cutting up the dried paper and counting in a scintallation
counter.

It is advisable to monitor the stability of triphosphates
periodically (as determined by their breakdown to mono- or
diphosphates--or even free phosphate) on storage, as well as
during or after an enzyme reaction.

E. The Specific Activity of Nucleotides

The specific activity of a triphosphate is usually expressed as the cpm per μmole. Assuming that the triphosphate
is radiochemically pure (see previous section) and contains no

free nucleoside or other material absorbing at A_{260} and A_{280},
the specific activity can be calculated, in neutral pH solution,
from the cpm per ml and the A_{260}. The molar extinction coefficients at 260 nm for ATP, CTP, GTP, and UTP are 15,300, 7,400,
11,700, and 10,200, respectively, at pH 7.0. Since the A_{280}/A_{260}
for these triphosphates are 0.15, 0.97, 0.66, and 0.38, respectively, the presence of other ultraviolet-absorbing materials
or another triphosphate (e.g., ATP) in a triphosphate preparation can be determined. As an example of calculating the ribonucleoside triphosphate specific activity, if a solution of
$[\alpha-^{32}P]UTP$ has an A_{250}/A_{260} of 0.74, an A_{280}/A_{260} of 0.38, and
the absorbance of the solution at 260 nm is 1.02, then it con-
tains 0.1 µmole of UTP per ml of solution. If the cpm, determined on a suitable dilution, of the same solution is 10^9 per
ml, then the specific activity of the UTP is 10^{10} cpm per µmole
of UTP, i.e., about 4.5 mCi per µmole of UTP.

III. REACTION MIXTURES FOR IN VITRO SYNTHESIS OF RNA

Reaction mixtures to measure the net synthesis of RNA
usually contain enzyme, RNA template, buffer, monovalent and
divalent cations, and nucleoside triphosphates--all optimized
in concentration to give the maximum level of RNA synthesis.
Additionally, to obtain maximum RNA synthesis, reaction mixtures
may require mercaptans, enzyme-protecting reagents [such as
bovine serum albumin (fraction V, Sigma)] and sometimes an
energy-generating system. The choice and optimization of

reaction conditions will be described below. The basic procedure for monitoring RNA synthesis is to incubate a suitable
reaction mixture containing a labeled triphosphate at a particular temperature and to determine the net synthesis of RNA
by measuring the incorporation of label into acid-insoluble
material.

A. A Typical Reaction Mixture

During preparation of the reaction mixture, the reaction
tube is kept in ice. Ingredients are added using separate
microliter pipettes (Bolab, Incorp., Derry, New Hampshire) or
disposable, 1-ml Plexiglass serological pipettes, and a disposable, Plexiglass, 13-mm diameter, culture tube. The value
of using disposable equipment is with regard to eliminating
cross-contamination of reagents, avoiding glassware or pipettes
that have been insufficiently washed (i.e., containing traces
of detergents or chromates, etc.), and for accuracy in pipetting.
The ingredients added are: distilled-deionized water--59μl; 1M,
sterile Tris·HCl buffer (pH 8.0)--8 μl; 1M, sterile, NaCl--8μl;
0.05 M ATP (pH 7.0)--2 μl; 0.05 M UTP (pH 7.0)--2 μl; 1 mM
$[\alpha-^{32}P]$UTP or $[^{3}H]$UTP (10 mCi per μmole)--10 μl; 1M, sterile,
$MgCl_2$--1 μl; 0.2 M dithiothreitol (DTT)--1 μl; enzyme (1 mg of
protein per ml)--10 μl; template (1mg of RNA per ml)--10 μl.

When the rate of reaction is determined, a ten-fold mixture without template is prepared and aliquots (115 μl) are
pipetted into separate reaction tubes, brought to reaction temp-

erature before addition of the template. A zero-time, unincu-
bated sample is removed and precipitated with two drops (about
0.1 ml) of a mixture of saturated sodium pyrophosphate and saturated Na_2HPO_4. This is followed by 1 ml of cold 5% (w/v) tri-
chloracetic acid (TCA). Additional sampling of the incubated
reaction mixture is performed as required, and similarly precipitated. After storage at $4°C$, for 10 min or more, the TCA-
insoluble labeled precipitates are collected on membrane filters
(Schleicher & Schuell Inc., Keene, N. H., 0.45-μm pore size) and
washed with 10 successive 5-ml portions of cold 5% TCA. The
filters are dried (to remove TCA and water) and the amount of
radioactivity is determined with a suitable counter. When the
product RNA is composed of small segments of RNA (e.g., oligonucleotides), it may be necessary to add a coprecipitant (e.g.,
50 μg of RNA or protein). Normally, however, the 20 μg of
enzyme and template are sufficient to coprecipitate the product
RNA. When [^{3}H]triphosphates are used to monitor product synthesis it is important to keep the mass of the acid-insoluble precipitate as low as possible in order to avoid quenching of the
^3H label and consequent lower counting efficiency.

B. Optimization of Reaction Conditions

To determine optimal reaction conditions of RNA synthesis,
several parameters of product synthesis have to be considered.
For example, the rate and duration of net RNA synthesis should
be determined. The integrity and relevance of the product to

the template RNA must be examined also. It is not sufficient
to set up a reaction mixture and determine the incorporation of
a labeled triphosphate into acid-insoluble material at two time-
points, e.g., 0. and 60 min. If, in such a reaction, the rate
of RNA synthesis is only linear for 10 min and then plateaus,
then the "observed" rate of reaction, based on the sampling at
60 min, would be actually one-sixth of the actual initial rate.
Hence, it is important to determine the incorporation of label
at several time points, and to plot the results as a function
of time.

    1.   Interpretation of Reaction Kinetics. Several guide-
lines for interpreting the types of kinetic data that are often
obtained are listed below (although they must serve only as an
initial guideline for the investigator).

        a.   The Reaction Rate Slows Down Then Terminates.
The enzyme may be denatured, oxidized, or digested by a protease--
this situation can be determined by addition of more enzyme to
the reaction mixture after the incorporation has ceased. Alternatively, the template may have been completely transcribed or
degraded, in which case the investigator should add more template
to see if this addition restimulates incorporation of triphosphate into RNA. The possibility of ribonuclease contamination
should be checked also by adding labeled RNA to the reaction mixture and monitoring its fate in terms of its integrity (by gel
electrophoresis) or degradation into acid-soluble material. It

is also possible that the substrates may be exhausted--as, for
instance, where there was not enough triphosphates or they were
degraded by a contaminating phosphatase. In such cases the integrity of the substrates should be monitored, and the situation
remedied by the addition of more of an "energy-generating system"
(see Section III.B.7). The pH of the reaction mixture should
be redetermined after the incubation period. Product inhibition
can also cause arrest of the reaction and of further product
synthesis.

   b.   The Reaction Rate Increases as a Function of Time.
The product acts as a template; with more RNA present, the enzyme
has more template to utilize. In such circumstances more of the
original template should be added to the reaction mixture.

   Alternatively, it is possible that the product is a better
template for the enzyme than the initial template added. If
so, the product RNA should be purified and added to a second
reaction mixture. Other RNA templates should be assayed.

   Another possible explanation is that an inhibitor is being
removed by one of the reaction constituents. For example, if
one of the presumed nucleoside triphosphates is actually its
corresponding diphosphate, and the reaction mixture contains
an energy-generating system, with the conversion back to its triphosphate the enzyme activity could then increase.

   c.   The Amount of Labeled Product Decreases. Usually
this indicates that ribonuclease is present and is degrading the

product RNA. The reagents, as well as template and enzyme, should be checked for contaminating ribonuclease, and purified by sucrose gradient centrifugation or by column chromatography.

        d. The Reaction Has an Initial Lag. The enzyme and/or template in such circumstances may be initially unavailable to each other (due to secondary structure, low concentration, etc.). Their concentration in the reaction mixture should be increased. Often a simple explanation of a reaction lag is that the reaction mixture temperature is initially below the water-bath temperature.

        While these are only general guidelines, it should be understood that much information can be gleaned from the kinetics of a reaction, and many of the problems encountered can be overcome through suitable experimentation.

        2. The Optimal Temperature for RNA Synthesis. Using a series of water baths set at different temperatures and an identical set of reaction mixtures, one may determine the optimal temperature for RNA synthesis. RNA polymerases vary in their stability and interactions with various substrates; consequently, it is not surprising to find that the optimal temperatures for their activities also vary. Kern Canyon, influenza, and VSV RNA transcriptases, when assayed in vitro, exhibit temperature optima around 31°C [2, 10]. For Qβ replicase the temperature optimum is between 37° and 39°C [11]. However, with reovirus the in vitro temperature optimum is between 45°C and 55°C [unpublished data].

3. The Monovalent Cation Requirement. The effect of
monovalent cations--such as lithium, sodium, potassium, and
ammonium--should be determined in reaction mixtures with regard
to their requirement, as well as to their optimum concentration.

4. Divalent or Other Cation Requirements. Similarly, the
effect of divalent and other cations should be determined both
singly and together. Influenza transcriptase responds best
in vitro in the presence of both manganese and magnesium ions.
However, VSV transcriptase is significantly inhibited in vitro
by the presence of manganese ions. Since nucleoside triphosphates
form salts with these cations, it should be remembered that if
the reaction mixture triphosphate concentrations are altered
significantly, then the divalent cation concentrations may also
have to be changed.

5. Optimization of the Nucleoside Triphosphate Reaction
Concentration. In order to conserve isotope, the labeled
triphosphate is frequently used in a reaction mixture at a lower
concentration than the other three unlabeled triphosphates.
However, before this is done, the reaction threshold concentration of each triphosphate should be determined (i.e., the minimum amount that must be present for the optimal reaction rate).
This threshold concentration may not be the same for each of
the four triphosphates and should, therefore, be determined for
each in a particular enzyme system.

6. Inclusion of Mercaptans. It is often desirable, in

order to prevent enzyme oxidation, to include a mercaptan in the
reaction mixture. One of the most useful is dithiothreitol (DTT),
and its optimal reaction concentration should be determined.
Since mercaptans have the property of keeping enzymes and other
reaction constituents in a reduced state, and since some enzymes
can be inactivated on reduction, alternative means for prevent-
ing oxidation (such as a nitrogen atmosphere, or covering the
reaction mixture with an inert mineral oil) should also be examined.

7. Inclusion of an Energy-Generating System. If the
nucleoside triphosphates in a reaction mixture are degraded during incubation by contaminating phosphatases, an "energy-generating system" can be included--e.g., pyruvate kinase and phos-
phoenol pyruvate. However, the amount included should be gauged
by the level of triphosphate degraded by the phosphatases. Both
with or without an energy-generating system, the fate of both
labeled and unlabeled triphosphates should be monitored during
the reaction. Since the maintenance of all four triphosphates
can be a complicated balance to obtain, and since there is a
risk of introducing contaminating nucleases with the pyruvate
kinase, a simpler, alternative approach is to add more triphosphate to the reaction mixture (provided the amount added is
not inhibitory and the magnesium optimum is redetermined).

8. Buffers, pH and Anions. Different buffers having
various pH values should be tried in reaction mixtures to

determine the optimum conditions for an enzyme reaction. Although Tris·HCl is a favorite buffer, others should be considered, particularly glycine buffers for high pH values, phosphate
buffers, etc. Since pH affects the charge and degree of ioniza-
tion of reaction constituents, this parameter should be one of
the first investigated. The pH of the buffer must be determined
at the reaction concentration and temperature of incubation, and
any change in pH during a reaction should be determined and corrected by increasing the amount of buffer.

The effect of sulfates, chlorides, phosphates, etc., on
an enzyme reaction should also be ascertained.

9. Detergents. Where an enzyme is enclosed in a membrane
system (such as in a virus particle), partial solubilization of
the membrane may be required before ions can reach the enzyme.
Various nonionic detergents (e.g., Triton N101 or Nonidet P40,
Shell Chemical Co.) and/or weakly anionic detergents (e.g.,
sodium deoxycholate) can be used. The optimum concentration
should be determined with each virus preparation.

10. Nuclease Inhibitors. When the amount of ribonuclease
in an enzyme preparation is sufficient to inhibit the synthesis
of RNA (and destroy the added RNA template), a nuclease inhibitor
such as bentonite or polyvinyl sulfate may be added. However,
since these additives often bind to, and inhibit, the enzyme they
should be used with caution.

11. The Addition of Ethylenediaminetetraacetic Acid (EDTA).
Although EDTA forms complexes with magnesium and removes it from

the reaction mixture, it is frequently added to a reaction mixture in low concentrations (e.g., 5% of the magnesium ion concentration). Triphosphates, NaCl, Tris·HCl buffers, magnesium
salts, the enzyme, and the template often contain trace amounts
of barium, iron, lead, etc., which are highly toxic to many
enzymes, and have a greater affinity for EDTA than magnesium.
By way of example, most triphosphates are prepared by a procedure involving barium precipitation, so that some barium ions
may remain in the final preparation.

12. Titration of the Amount of Enzyme or Template, and
Addition of Protein Stabilizers. The linear response of the
reaction rate to the amount of enzyme should be determined.
Since too much protein can inhibit RNA synthesis, reactions
should be run within the permissible protein concentrations.
Too little enzyme protein can result in loss of activity due to
enzyme inactivation (e.g., oxidation). This situation can frequently be remedied by including in the reaction mixture, or
enzyme stock, bovine serum albumin (BSA: fraction V, Sigma).
The amount of BSA added must be optimized, in case its addition
is also inhibitory. The optimum amount of RNA template for
maximum product synthesis should be ascertained.

C. The Integrity of the RNA Synthesized as Witnessed by
the Reaction Requirements

When an enzyme is templated by RNA and synthesizes complementary RNA product, then it should require the presence of
all four ribonucleoside triphosphates in the reaction mixture.

If one or more triphosphates is omitted from the reaction mixture and RNA is being synthesized--as monitored by the incorporation of another labeled triphosphate--then it is clear that the product copy cannot be complementary to the original RNA template molecule. In such a case the first reaction parameter that must be checked is the purity of the triphosphates present. Since labeled triphosphates are usually synthesized with ATP as a donor (as described above), possible ATP contamination might be suspected. Moreover, since CTP can be relatively easily deaminated to UTP, this can be another source of triphosphate contamination. However, these possibilities can be determined easily by Dowex-1-Cl⁻ chromatography or by high-voltage paper electrophoresis.

Where triphosphate contamination is precluded, then one possible explanation is that the enzyme (or a contaminant) is adding nucleotides covalently to the end of a tamplate molecule. In these situations the reaction may have no template specificity, and any RNA homo- or heteropolymer will stimulate triphosphate incorporation. If the enzyme uses only one homopolymer as template (and one complementary triphosphate) this fact may give the investigator a useful tool for the purification of that specific enzyme activity and allow him to separate it from other proteins--even proteins with which it may have been originally associated. By way of example, the viral-specified Qβ protein of Qβ replicase uses poly(C) as a template and makes poly(G)

product (therefore, GTP is the only triphosphate required). This
fact has allowed the separation and purification of this protein
from the other protein components of the composite Qβ replicase
[12].

If only one or two triphosphates are required for RNA
synthesis in a reaction mixture, nearest neighbor analysis of
the product will allow the investigator to determine if a specific homo- or heteropolymer is being synthesized or whether the
triphosphates are being added one per 3' end of the template RNA
molecules (see Section IV.D.3).

D. Calculation of the Rate of RNA Synthesis from the
Incorporation of a Labeled Nucleotide

To determine net RNA synthesis, the specific activity of
the incorporated nucleotide must be known. If the labeled
nucleoside triphosphate remains intact in the reaction mixture,
the rate of incorporation can be calculated from the net incorporation (e.g., the 60 min amount minus the zero-time "blank"),
divided by the specific activity of the triphosphate. Thus, a
reaction containing 0.01 μmole of $[\alpha-{}^{32}P]UTP$ (1×10^{10} cpm per
μmole) and 0.09 μmole of unlabeled UTP was incubated with the
other reaction constituents. The zero-time sample gave a "blank"
incorporation of 100 cpm and the 20-min, 40-min, and 60-min
samples gave incorporations of 3400 cpm, 6700 cpm, and 10,000
cpm, respectively. The UTP specific activity in the reaction
mixture was $(0.01 \times 1 \times 10^{10})/(0.09 + 0.01)$ cpm per μmole =
1×10^9 cpm per μmole (1×10^3 cpm per pmole). Therefore, the

rate of incorporation was $\dfrac{10,000-100}{10^3}$ pmoles, i.e., about 10 pmoles

of UMP per hr, and the rate of incorporation was linear through

the first hour of incubation. If the reaction contained 10 µg

of enzyme protein, then this rate of incorporation was 10 x 100,

i.e., 1000 pmoles of UMP per mg of protein per hr.

If all four nucleotides (AMP, GMP, CMP, and UMP) are incorporated equally into the product RNA (as determined by nearest neighbor analysis, or from the base ratio of the product RNA
or by double triphosphate incorporation experiments, III.E),
then the net rate of RNA synthesis can be calculated. The
weights of 1 pmole of AMP, CMP, GMP, and UMP (as the free acids) are
0.347, 0.323, 0.363, and 0.324 pg, respectively, or, by sum,
1.357 pg for all four. Since during the snythesis of RNA an
H_2O molecule is excluded per nucleotide addition, the sum RNA
weight for all four nucleotides incorporated is 1.285 pg, about
1.3 pg. Assuming that all four nucleotides are incorporated
equally, then for every 1 pmole of UMP incorporated there is 1.3
pg of RNA synthesized. In the example given in the previous
section (1000 pmole of UMP incorporated per mg of protein per
hr), the net rate of RNA synthesis (calculated as the free acid)
is 1300 pg (0.0013 µg) per mg of protein per hr. If the reaction
was templated by 10 µg of RNA (i.e., 10 x 100 µg of RNA per mg
of protein), this net rate of product synthesis would constitute
a net increase in RNA of $\dfrac{0.0013 \times 100\%}{1000}$ = 0.00013% net RNA increase.

E.   Incorporation of two Labeled Nucleoside Triphosphates

In order to determine the extent of incorporation of each
of the ribonucleotides, double-labeling experiments may be used.
With a ^{3}H-labeled triphosphate (e.g., [^{3}H]UTP), the incorporation of the other three [α-^{32}P]ribonucleoside triphosphates can
be determined in three separate reaction mixtures, and the rate
of [^{3}H]UMP incorporation in each used to normalize the data
between the reactions. The relative rates of incorporation of
the two labeled triphosphates should be compared in these kinetic experiments to see if there is preferential utilization of
a triphosphate as a function of time.

F.   Initiation of RNA Synthesis

RNA-dependent RNA polymerases initially synthesize complementary product RNA, starting at the 3' end of the template
molecule. Product RNA synthesis is started by condensing two
triphosphates together. The three phosphates of the terminal
nucleoside remain on the product strand, although the penultinate
nucleoside loses its γ- and β-phosphates and retains its α-phos-
phate in a phosphodiester link with the 3' hydroxide group of
the ribose of the terminal nucleotide. Therefore, in the synthesis of pppApG.... initial 5' product strand sequence, the β
and γ phosphates of GTP are lost.

All succeeding additions of nucleotides similarly involve
loss of the β-, and γ-phosphates of the added nucleotide, so
that the 5' terminal nucleotide is unique on the product strand

by its retention of its three phosphates. By including $[\gamma-^{32}P]$- or $[\beta, \gamma-^{32}P]$ribonucleoside triphosphates in a reaction mixture, specific labeling of the 5' terminus of the product molecule can be obtained. If the 5' nucleotide is GTP, only $[\gamma-^{32}P]$-, or $[\beta, \gamma-^{32}P]$GTP of the four similarly labeled triphosphates will label the product RNA.

Alkali digestion of RNA cleaves the polynucleotide to give nucleoside, 3' (or 2', or 2': 3' cyclic) monophosphates-- except for the terminal nucleotide, which gives a nucleoside tetraphosphate (e.g., pppAp in the preceding example). The nucleoside tetraphosphates can be readily resolved from mono-, di-, or triphosphates by paper electrophoresis, DEAE-cellulose-urea column chromatography [13], or two-dimensional thin-layer chromatography [14].

When the 5' terminal nucleotide of the product RNA species is known, the average molecular weight of the product RNA can be determined by including in the reaction mixture a $[\beta, \gamma-^{32}P]$- or $[\gamma-^{32}P]$-triphosphate (to label the 5' terminus of the RNA), and a base-labeled (with ^3H) triphosphate (to monitor total RNA synthesis). If the base-ratio of the product is known, then the average molecular weight of the product can be determined from the relative incorporation of the two labels after correction for their respective specific activities and the base-ratio of the product.

Even if only one of the four $[\beta, \gamma-^{32}P]$- or $[\gamma-^{32}P]$tri-phosphates labels the product RNA, it is important to confirm

its location at the 5' terminus of the product strand. Apart
from alkali hydrolysis to retrieve the label as a ribonucleoside
tetraphosphate, ribonuclease digestion can be used to identify
the 5' (oligo)nucleotide. When the 5' nucleotide is GTP, digestion by ribonuclease T_1 will give guanosine tetraphosphate
(pppGp). When the 5' nucleotide is CTP or UTP, ribonuclease A
digestion will give the corresponding cytidine or uridine tetraphosphates. However, if in the latter case, ribonuclease T_1,
or, in the former case, ribonuclease A is used for the digestion,
the 5' terminal oligonucleotide can be obtained and isolated by
DEAE-cellulose-urea column chromatography or two-dimensional
paper electrophoresis.

Once the 5' terminal nucleotide has been identified,
attempts can be made to determine its nucleotide neighbor by
determining which of the four $[\alpha-^{32}P]$ribonucleoside triphosphates
labels the ribonucleoside tetraphosphate. By including all four
labeled $[\alpha-^{32}P]$ribonucleoside triphosphates in a reaction mixture and isolating the 5' terminal oligonucleotide, one can
obtain the sequence of the oligonucleotide by specific nuclease
or phosphodiesterase digestions (see Section IV.D.1.).

G. Analysis of Reaction Mixtures with Labeled Template RNA

In examining RNA synthesis by RNA-dependent RNA polymerases, the investigator is frequently concerned about the properties of the reaction product RNA. Equal concern should be

given to the fate of the template molecules. The involvement of template RNA with enzyme molecules, initiation of product synthesis, and the involvement of product molecules with template RNA in replicating complexes are all important parts of the reaction being studied. If the template RNA is labeled, its fate through a reaction can be followed. Where an alternative radioisotope is used for the ribonucleoside triphosphates, both template and product molecules can be monitored.

Two added advantages of including labeled template RNA in a reaction mixture should be mentioned. One is to monitor nuclease contamination of the reaction mixture. If the labeled template RNA, with or without the occurrence of product synthesis, is digested to acid-soluble nucleotides or degraded to small pieces by the enzyme preparation, nuclease activity is present. When nuclease contamination is precluded, the recovery of acid-insoluble template label during a reaction can be used to normalize sampling variations. If a reaction mixture contains 10^4 cpm of ^{3}H-labeled template RNA per 0.1 ml, and the rate of product synthesis is being ascertained by the incorporation of an $[\alpha\text{-}^{32}P]$ribonucleoside triphosphate into RNA, variations in the 10^4 cpm of $[^3H]$RNA (recovered per 0.1-ml sample) can be used to correct the observed acid-insoluble ^{32}P counts, and thereby give a "normalized" ^{32}P incorporation. This correction is particularly valuable where the template, enzyme, and product are present in particles (such as membranes or

virus aggregates) that precipitate during the reaction. Similarly, when "duplicate" reactions are being run, templated by the
same amount of RNA (or virus), errors in the amount of RNA (or
virus) added can be determined and, where this is a factor in
the amount of resulting product RNA synthesis, a correction can
be made.

IV.   ANALYSIS OF THE REACTION-PRODUCT RNA SPECIES

The feasibility of having both template and product labeled depends on the system being studied. However, since it is
desirable to follow the fate and interrelationship of both template and product molecules, the succeeding discussion will describe experiments designed to investigate both RNA species simultaneously.

A.   Purification of RNA from a Reaction Mixture

Protein and labeled triphosphates can be removed from RNA
by phenol extraction followed by Sephadex G-50 column chromatography. A 1-ml reaction mixture is mixed into 0.1 ml of 10% (w/v)
sodium dodecyl sulfate (SDS), and 0.02 ml of E. coli ribosomal
RNA (1 mg/ml,--or another RNA, to act as a carrier during the
alcohol precipitation) followed by 0.5 ml of a phenol:m-
cresol:8-hydroxyquinoline mixture (500:70:0.5 w/w/w). After
shaking for 1 min, the phases are separated by centrifugation
at 1000 X g for 5 min, and the aqueous phase is mixed with 0.1
ml of glycerol; the RNA is separated from triphosphates by passage through a 90 x 1 cm column of Sephadex G-50 (coarse grade),

equilibrated with 0.4 M NaCl, 0.01 M Tris·HCl (pH 7.5), 5mM
EDTA, and 0.1% SDS. The column eluant is monitored for radioactivity, and the excluded volume, containing RNA, is mixed
with two volumes of ethanol and stored in a siliconized
(Siliclad, Scientific Products, Evanston, Illinois) Corex centrifuge tube (Corning, Corning, N. Y.) at -20°C for 4 hr. Siliconi-
zation aids the recovery of RNA after centrifugation, since in
such tubes the RNA is recovered at the bottom, rather than down
the side of, the tube. The RNA is retrieved by centrifugation
at 20,000 X g for 30 min at -4°C in an HB4 swinging bucket rotor
(Sorvall, Norwalk, Conn.), and is then dissolved in 1 ml of 0.4
M NaCl, 0.01 M Tris·HCl (pH 7.4) and precipitated with two volumes of ethanol to remove residual traces of SDS. After a
second similar centrifugation, the RNA is dissolved in a minimum volume (e.g., 0.2 ml) of 5 mM EDTA, 0.01 M sodium phosphate
(pH 7.0) and stored frozen at -20°C.

B. Polyacrylamide Gel Electrophoresis of RNA

Due to their content of phosphate, RNA molecules move in
an electric field (at pH 7) toward the anode. When a molecular
sieve (such as polyacrylamide or agarose gel) is placed between
the RNA sample and the anode, RNA species may be resolved according to their mass--provided that the gel pores are large enough
to accommodate all species of RNA present. During electrophoresis there is an approximately linear relationship between the
log of the molecular weight of an RNA molecule and the relative
distance it migrates in a particular gel [15]. Presumably, this

reflects the ability of small RNA molecules to pass freely
through both small and large pores of the gel, while bigger
molecules are delayed by the small pores and only pass through
the larger ones.

Polyacrylamide gel electrophoresis provides a useful tool
for examining the relationship of template and product RNA
species. When newly synthesized product RNA is present among
the reaction product nucleic acids, hydrogen-bonded to a template molecule, it forms a complex that is larger and, upon
electrophoresis, slower moving, than the free template molecule.
The amount of template recovered in these complexes reflects the
number of template molecules involved in enzyme activity. From
the relative specific activities of template and product RNA
species, the relative masses of product and template RNA in
these complexes can be determined.

Where product RNA species are also present in a reaction
mixture as free molecules, the relative amounts of free and
complexed product RNA can be ascertained from the distribution
of product label through the gel after electrophoresis. Similarly, where template RNA is degraded during a reaction by contaminating nucleases, the amount of degradation can be monitored
by the amount of low molecular weight template species generated
during the reaction. Therefore, several important parameters of
the RNA polymerase reaction can be studied simultaneously by
gel electrophoresis of purified reaction product nucleic acids.

(See also Chapter 12 of Volume I of this series).

1. Preparation of Polyacrylamide Gels. In view of the
quenching of ^{3}H counts by dried polyacrylamide gel slices, two
types of gel can be used: a bis-acrylamide cross-linked gel--
which can be dissolved by hydrogen peroxide--and an ethylene
diacrylate cross-linked gel--which can be dissolved by alkali.
The lowest acrylamide concentrations yielding gels that can be
readily manipulated are 1.8% for gels cross-linked with bis-
acrylamide and 2.4% for those in which ethylene diacrylate is
used.

Acrylamide and bis-acrylamide can be recrystallized from
chloroform and acetone, respectively. However, the materials
supplied by BioRad Laboratories (Rockville Center, N. Y.) are
superior to any we have purified. Ethylene diacrylate is obtained from the Borden Chemical Company (Philadelphia, Pa.).

Polyacrylamide 2.4% gels are prepared by mixing 4 ml of
an aqueous stock solution of 15% recrystallized acrylamide--0.75%
recrystallized bis-acrylamide with 12.45 ml water and 8.33 ml of
3E buffer (0.12 M Tris, 0.06 M sodium acetate, 3 mM sodium EDTA--
adjusted to pH 7.2 with 6 ml of glacial acetic acid). Air is
removed from the mixture by evacuation and 0.02 ml of N, N, N',
N'-tetramethylethylenediamine (TEMED) and 0.2 ml of fresh aqueous 10% ammonium persulfate solution are added. After swirling,
the mixture is transferred to Plexiglass tubes (0.9 cm internal
diameter) with rubber stoppers in one end, and allowed to polymerize for 40 min in an upright position. The gel length is

exactly 10 cm. Due to extrusion of liquid during polymerization, the gel surface is flat. After polymerization, the gels are transferred to 500 ml of E buffer (one-third the concentration of 3E) containing 0.1% recrystallized sodium dodecyl sulfate (SDS). The gels are left for at least 72 hr in order to remove materials that absorb ultraviolet light, and also to swell to 150% of their original length (180% of their original volume). No further swelling is obtained if the gels are left for longer periods. Swollen gels can be cut in half prior to their being taken back into the electrophoresis tube; in such cases the cut end of the gel is uppermost and is used for loading the RNA sample. When gels are to be scanned to locate RNA using ultraviolet light at 260 to 280 nm, it is preferable to use swollen gels prepared by the above method. However, where only radioactive RNA is being analyzed the swelling process is unnecessary and the gels can be used directly.

Diacrylate gels are prepared similarly using a stock solution of 15% recrystallized acrylamide and 1.0% ethylene diacrylate. For gels having an acrylamide concentration greater than 4%, it is necessary to suck the gel up and down the tube gently after polymerization has commenced in order to prevent gel adherence to the tube walls--this movement facilitates the final extrusion of the gel. At lower acrylamide concentrations, and in contrast to glass tubes, the gels slip easily out of the Plexiglas tubes.

2. Electrophoretic Conditions. The electrophoresis buffer we use is buffer E containing 0.1% SDS [15]. Swollen gels are

sucked into 0.9-cm (internal diameter) Plexiglass tubes and
supported by a dialysis membrane stretched across the lower end.
They are then prerun at room temperature for 30 min at 10 mA per
tube and 50 V (5.5 V/cm); a Buchler, or similar voltage- and
current-regulated dc power supply, set for current regulation,
is used. Unswollen gels must be prerun for 2 hr or more to
allow SDS to migrate into the gel. The distance that SDS has
migrated can be ascertained by cooling the gel to 4°C; the SDS
crystallizes and can be visualized.

Samples of nucleic acid are added to the gel in 10%
glycerol in the minimum loading volume possible in order to
obtain good resolution. Excellent resolution is obtained with
less than 25 μl; samples of 200 μl can be used, but yield a
poorer resolution. The maximum amount of RNA that can be applied to one gel of the size specified is between 200 and 400 μg,
although the best resolution is obtained with less than 20 μg.
Electrophoresis is done at room temperature, 10 mA per tube and
50 V for 90 min or more, as required.

3. Determination of the Location of RNA after Electrophoresis. After electrophoresis, the gels can be scanned by
transfer to a rectangular quartz cell (1.0 cm X 12.0 cm X 2.0cm)
and scanned by transmitted ultraviolet light in a Joyce high-
resolution Chromoscan with a 266-nm interference filter. At the
end of the scan the length of the gel in the cell must be
measured, and subsequently remeasured after floating in water on

a right-angle aluminum trough in order to obtain the natural
gel length. Water is then decanted and the gel is frozen in
the aluminum trough on powdered dry ice. After freezing, each
gel is transferred to a carbon dioxide gas-cooled microtome and
sequentially sliced; the 0.5-mm slices are dried on filter paper
or placed in scintillation vials. Slicing is facilitated by the
presence of sodium dodecyl sulfate in the gel, as the sodium
dodecyl sulfate precipitates on cooling and inhibits the formation of large ice crystals. The resolution obtained between
different RNA species is evidence of the fact that this method
of freezing does not distort the RNA bands in the gel.

Gel slices containing only ^{32}P and dried on filter papers
can be counted in toluene-based BBOT (2,5-bis [2(5-tert)butylben-
zoxazolyl)] - thiophane) with a Packard or other liquid scintillation spectrometer. Diacrylate gel slices are dissolved in 0.5
ml of a mixture of alcoholic Hyamine hydrozide and 1 M piperidine
(1:9 v/v) before counting in Kinard's scintillation fluid or
Aquasol (New England Nuclear Co., Boston, Mass.).

Slices of bis-acrylamide cross-linked polyacrylamide (1 mm)
gels can be dissolved in 0.5 ml of "stabilized" 30%(v/v) hydrogen peroxide by incubating them at 60°C in a sealed scintillation vial until they are dissolved. The solution is then cooled
and mixed with Kinard's scintillation fluid, Aquasol, or a
similar scintillation cocktail in which aqueous material is
soluble. A mixture of 1 part of Triton X100 and 3 parts of

BBOT-xylene or toluene is convenient and inexpensive for this purpose. After correction for the isotope counting efficiencies and the crossover of ^{32}P label into ^{3}H or ^{14}C counting channels, the amount of labeled RNA in each gel slice can be ascertained and the RNA profile obtained and related to any ultraviolet scan performed.

C. Separating, Melting, Annealing, and Ribonuclease Digestion of Reaction-Product Nucleic Acids

To establish the complementary or identical nature of product RNA to that of the progenitor template species, hybridization experiments can be performed. Since the total reaction-product nucleic acids contain both product and template species, a small portion of which may be complexed together, product and template species must be separated from each other, or the hybridization experiments must be designed with these facts in mind. When product RNA species are smaller than the template molecules, column chromatography, sedimentation, or gel electrophoresis and elution of the gel slices can be used to separate product from template. The method chosen should relate to the degree of resolution desired and the amount of material to be processed. Gel electrophoresis is excellent for amounts of nucleic acids below 100 µg, sedimentation for amounts under 1 mg, and columns can be designed to handle gram quantities. Since low molecular weight RNA is included in the gel beads of columns (see below), and the volume in which it elutes is large, this method is not recommended for small quantities of material.

1. Elution of RNA from Gel Slices. Procedures for separating unmelted or melted (see below) nucleic acids by sedimentation will not be described here since this technique is
very well known. Electrophoresis of melted or unmelted RNA can
be conducted as described above. Subsequent elution from the
gel slices can be performed by one of two methods. The gel
slices can be eluted with 0.4 M NaCl, 0.01 Tris·HCl (pH 7.4),
0.01%(w/v) SDS at 4°C for 24 hr. Alternatively, the slices can
be eluted at 20°C with 50% v/v Me_2SO containing 5 mM EDTA buffered to pH 7. This latter method is very efficient and fast
for most RNA species studied and is used routinely by us when
maintenance of secondary structure (double-strandedness, etc.)
is not important. In either case the RNA elution is monitored
by the release of label from the gel slice. RNA is purified by
filtration or centrifugation to remove polyacrylamide, then precipitated with alcohol and recovered by centrifugation. Where
Me_2SO is present, the eluant solution is diluted five-fold
with 0.4 M NaCl before precipitation with 3 volumes of alcohol.
It is strongly advised to use preswollen gels for the electrophoresis in order to reduce the amount of free acrylamide and
undesirable ions in the gel slice eluants.

2. Exclusion Column Chromatography. Exclusion column
chromatography can be used for the concomitant isolation of
product-template complexes. For example, VSV product-template
complexes from free VSV viral (template) RNA and free VSV transcription product RNA (small molecular weight) can be purified by

column chromatography on 1% or 2% agarose. A column (90 cm X 1

cm) is prepared containing 1% agarose--suspended in, and eluted

by, 0.4 M NaCl, 0.1 M Tris·HCl pH 7.4, 3 mM EDTA, and 0.1% SDS.

The reaction product nucleic acids are loaded on the column in

10% (w/v) glycerol--after phenol-cresol-SDS extraction--and the

column eluants are monitored for both template [^{3}H] RNA and [^{32}P]-

product species. The product-template complexes are recovered

in the excluded volume, free VSV viral RNA is recovered in the

fractions that come just after the void volume, while free VSV

transcriptase product RNA is retrieved in subsequent included

fractions.

   3. Synthesis and Separation of Brominated Product RNA by

Density Differences. Where native product RNA cannot be obtained

free from template species by sieving, electrophoresis, or sedi-

mentation, and the amount of total product is less than the

amount of total template present, the only alternate approach

for separating the two species is exploiting potential density

differences. By including brominated UTP (Br-UTP) or CTP (Br-

CTP) in the reaction mixture, product RNA can be made that is

heavier--due to its bromine content--than the template species.

After melting, the brominated species can be separated from non-

brominated RNA by CsCl density equilibrium centrifugation.

   4. Denaturation of RNA Complexes. Denaturation of RNA

complexes using heat or denaturing solvents, such as Me$_2$SO or

formamide, must be monitored not only with regard to the efficacy

of denaturation (i.e., loss of complexes or ribonuclease resistance) but also with regard to the integrity of the free RNA on
exposure to the melting agent. If, in denaturing the complexes,
the process degrades the RNA, then the procedure has to be re-
vised. Although melting of RNA species is not concentration-
dependent, reannealing of the melted species is; this can be a
cause for frustration where the success of the melting process
is thwarted by subsequent reannealing [16]. We have found that
melting an RNA sample in 0.2 ml of 0.01 M sodium phosphate buffer
(pH 7.0) containing 5 mM EDTA at 100°C for 20 sec is usually
successful. However, for double-stranded RNA of poliovirus, the
concentration of RNA must not exceed 1 µg per ml or reannealing
takes place [16]. For Qβ, VSV, or influenza double- or multistranded RNA, melting can be accomplished successfully at higher
RNA concentrations (10 to 50 µg per ml) [17,18]. Since divalent
cations (such as magnesium ions) catalyze the degradation of
RNA, EDTA is included in the buffer. Such cations are also a
cause for concern when using denaturing solvents such as Me_2SO or
formamide. These solvents, which lower the melting temperature
of double-stranded molecules, should be distilled before use,
and used in conjunction with EDTA.

    5. Annealing RNA Species. Annealing of RNA depends on
concentration, time, salt, and temperature. These parameters
should be determined to optimize the annealing conditions for a
particular RNA. Normally, we perform RNA-RNA annealing in

aqueous solution at a concentration of RNA of 10 µg/ml with

respect to one of the species (usually the template moiety) in

0.4 M NaCl, 0.01 M Tris·HCl (or sodium phosphate) buffer (pH

7.0), 5 mM EDTA to 60°C for the desired length of time (under

these conditions, 30 - 60 min is usually sufficient). Denatur-

ing solvents can also be included and the annealing temperature

lowered.

The success of the annealing process can be judged by two

criteria: generation of ribonuclease-resistant duplexes and

formation of RNA complexes. Conversely, in the melting proce-

dure its success can be judged by the loss of these attributes.

Double-stranded RNA is stabilized by the presence of salts (NaCl,

KCl, or LiCl, for example), due to their reduction of the inher-

ent electrostatic repulsion between the phosphodiester backbones

of the two RNA strands--and, therefore, indirect net enhancement

of the hydrogen-bonding of base pairs in the structure. Ribo-

nuclease digestion of RNA, as a measure of the amount of double-

strandedness, should, therefore, be performed in a buffer solu-

tion containing salt.

6. Ribonuclease Digestion of RNA. We usually digest RNA

in 0.4 M NaCl, 0.01 Tris·HCl (or sodium phosphate) buffer (pH

7.0), 5 mM EDTA, and use ribonuclease A and T_1 (10 µg of each

per ml). Since these enzymes digest single-stranded RNA to give

oligonucleotides terminating with uridine and cytidine 2' or 3'

(or 2': 3' cyclic) monophosphates (for RNase A) or guanosine

2', 3' monophosphates (for RNase T_1), theoretically all single-
stranded RNA regions except those involving adenosine residues
should be digested. The amount of polyadenylate residues as
well as the amount of nuclease-inaccessible regions of the RNA--
due to secondary structure and intrastrand hydrogen-bonding--
constitute the "core" of acid-insoluble, ribonuclease-resistant
material obtained by digestion of single-stranded RNA. The difference between the ribonuclease-resistant "core" value of
single-stranded RNA (e.g., an unannealed sample) and that obtained by annealing complementary RNA strands represents the
amount of double-stranded RNA generated by the annealing process.

If annealing is monitored by the generation of template and
product ribonuclease resistance as a function of time, then--
under conditions of template excess--the amount of complementary
product can be ascertained. This amount can be operationally
defined as the maximum amount of increase in product ribonuclease-
resistance obtained by annealing in template excess. The importance of having an excess of template (plus) strands is to
prevent, by competition, any product (plus) strands from annealing to complementary product (minus) strand RNA. In order to
further prevent product-product annealing, the product should
be diluted to such a low concentration that it does not have
the chance to anneal during the annealing incubation [10].

A frequently used procedure to determine if any product
plus strands exist among the spectrum of product RNA is to

self-anneal the product RNA in the absence of template nucleic
acid. Even where template contamination is rigorously excluded,
unless the amount of self-annealing is substantial, an alternative approach to the simple demonstration of ribonuclease-
resistance should be used. For example, new product-product
complexes should be demonstrated.

Annealing product and template (or product self-annealing)
will generate double-and multistranded complexes, which on gel
electrophoresis move slower than their free constituents. Resolution of annealed RNA species upon polyacrylamide gels therefore
provides a useful additional tool for the demonstration of
hybridization [17].

D. Nearest Neighbor- and Product-Sequence Analyses

RNA synthesis involves the incorporation of ribonucleoside
5' triphosphates into RNA. Alkali digestion of RNA releases
ribonucleoside 2' and 3' monophosphates. Therefore, the 2' and
3' monophosphates of a nucleoside obtained by alkali degradation
of RNA originate from its neighbor's progenitor triphosphate.
This nearest neighbor transfer of phosphate is a useful tool
for RNA sequence studies.

1. The Sequence of the 5' End of Product RNA. If the
sequence at the 5' end of a product RNA is pppGpUpAp.... etc.,
both $[\alpha-^{32}P]GTP$ and $[\alpha-^{32}P]UTP$ will label the guanosine tetraphosphate obtained by alkali (0.3 M NaOH, 37°C for 18 hr) or
ribonuclease T_1 digestion of the product RNA. However, the

dinucleotide pppGpUp obtained by pancreatic ribonuclease A di-
gestion of the same RNA will be labeled by [α-^{32}P]GTP, [α-^{32}P]UTP,
and [α-^{32}P]ATP. Confirmation of the nucleotide sequence can be
obtained not only by alkali digestion of the oligonucleotide
(whereby the [α-^{32}P]ATP label is recovered in UMP, and [α-^{32}P]UTP
and [α-^{32}P]GTP label are recovered in G-tetraphosphate) but also
by the use of phosphatases to remove the exposed 3' and 5' phosphates of the dinucleotide and leave the internal phosphate intact in the dinucleotide monophosphate (GpU). Since only
[α-^{32}P]UTP labels this dinucleotide (which after treatment with
alkali gives labeled GMP), the sequence can be obtained and
confirmed by various approaches [18].

     2. DEAE-Cellulose-Urea Column Chromatography of Nucleo-
tides. One method for separating oligonucleotides--particularly
the 5' and 3' terminal oligonucleotides generated by nuclease
digestion of product RNA--is that of DEAE-cellulose-urea column
chromatography [18]. In the presence of urea (to prevent oligonucleotide interactions) oligonucleotides can be eluted by a
salt gradient from DEAE-cellulose essentially according to
their phosphate charge at pH 7.8. The elution of oligonucleotides from DEAE-cellulose gives a series of isopleths whereby for
a ribonuclease A digest of a RNA, all nucleotides of the same
phosphate charge elute together. Therefore GpUp, ApUp, GpCp,
and ApCp are eluted from the column in the same peak. Resolution through the dodecyl oligonucleotides can be achieved by

careful chromatography. The methodology used is described
below for the separation and purification of ribonuclease A-
derived oligonucleotides.

DEAE-cellulose (Bio-Rad 0.95 meq/g) is freed from fine
particles by suspending in distilled water (200 g/10 liters) and
decanting the supernatant; it is then adjusted to pH 2 with HCl
(4 liters) and diluted to 10 liters with water. The supernatant
again is decanted. The DEAE-cellulose is washed twice with 10
liters of water, made 2 M with respect to ammonium carbonate
(4 liters total volume), and diluted subsequently to 10 liters
with water; the supernatant again is decanted. After two additional washes with 10-liter volumes of water, the slurry is
made up to 2 M NaCl (4 liters total volume), diluted to 0.2 M
NaCl, and the supernatant again is decanted. The slurry is
finally made up in 7 M urea, 3 mM EDTA, 0.01 M Tris·HCl, and
0.2 M NaCl (pH 7.8). Columns (27 X 0.8 cm) are supported on
sintered-glass disks and packed under 5 psi. Before use, a
column is washed with 100 ml of 7 M urea, 3 mM EDTA, and 0.01 M
Tris·HCl (pH 7.8) at a rate of 0.5 ml/min, regulated by a peristaltic pump.

RNA is digested with RNase A in 0.05 M EDTA, 0.10 M NaCl,
0.01 M Tris·HCl (pH 7.4) by mixture of an RNA sample with 11 mg
of E. coli RNA and 0.5 mg of pancreatic ribonuclease and incubation of the 1-ml mixture at 37° for 30 min. The digest is diluted ten-fold with 7 M urea and loaded directly on a DEAE-
celulose column and eluted with a 400-ml linear gradient of 0 to

0.35 M NaCl in 7 M urea, 3 mM EDTA, and 0.01 M Tris·HCl (pH 7.8).

The isolation of an isopleth from a DEAE-cellulose column is as follows: The pooled fractions are diluted six-fold with water and passed through a 9 X 0.5 cm column packed under 5 psi and previously washed with 50 ml of 0.05 M triethylamine bicarbonate buffer (pH 7.4). The oligonucleotides, which are adsorbed, are freed from residual urea and NaCl by a subsequent wash of the columns with 30 ml of 0.05 M triethylamine bicarbonate buffer, then eluted with 2 M triethylamine bicarbonate (pH 7.4). Finally, the triethylamine bicarbonate is removed by distillation with water under reduced pressure at 40°C. The oligonucleotides are converted into their sodium salts by Dowex 50 (H^+) treatment and neutralization of the supernatant with 0.1 N NaOH.

A 5' terminal oligonucleotide such as pppGpGpApUp has a net phosphate charge of 9 and should elute from DEAE-cellulose-urea with the octanucleotides $(Xp)_7Xp$. After digestion with alkaline phosphatase the oligonucleotide GpGpApU is obtained, which has a net phosphate charge of 3, whereas the octanucleotides minus their 3' phosphates have a net charge of 7 (i.e., $(Xp)_7X$). Rechromatography after phosphatase digestion, therefore, will resolve the 5' derived nucleotide from other nucleotides. Thereafter, sequence analysis by alkali, ribonuclease T_1 or spleen phosphodiesterase digestion will give the actual sequence of the original 5' oligonucleotide.

Nearest neighbor analysis with individually labeled $[\alpha-^{32}P]$ribonucleoside triphosphates should also give the 3' neighbor nucleoside to this 5' terminal nucleotide. If the sequence is pppGpGpApUp(U), the 5' terminal oligonucleotide will be can be used to demonstrate this fact.

3. The Sequence at the 3' End of Product RNA. The 3' terminal oligonucleotide of an RNA--such as ApApCpApApA-OH obtained by ribonuclease T_1 digestion of an RNA--will have a net charge of 5 and elute from DEAE-cellulose-urea with the tetranucleotides [i.e., $(Xp)_3Xp$]. Phosphatase digestion will convert tetranucleotides to a residue of net 3 phosphate charge [i.e., $(Xp)_3X$], but will leave the 3' oligonucleotide intact. Rechromatography can then be used to isolate it free from the other oligonucleotides.

A disadvantage of DEAE-cellulose-urea chromatography is encountered with the resolution of ribonuclease T_1 RNA digests. For T_1 digests, resolution of nucleotides into isopleths based on their net phosphate charge deteriorates for nucleotides greater than the penta- and hexanucleotides.

Other procedures, such as 2-dimensional thin-layer [14] or paper electrophoresis [19], or polyacrylamide gel electrophoresis in high concentration gels can also be used to separate nucleotides.

V. KNOWN RNA-DEPENDENT RNA POLYMERASES

The replication process of all RNA viruses so far studied--other than the oncornaviruses (e.g., Rous sarcoma virus)--apears to involve an RNA replication stage through some form of multi-

stranded RNA intermediate. In some cases the enzyme or RNA-
enzyme complex responsible for the replication process has been
identified and studied in detail (e.g., Qβ replicase). In most
other cases double- or multistranded RNA species have been found
in virus-infected cells and shown by hybridization to be homologous to the infecting viral genome. In the latter situation,
the presence of an RNA-dependent RNA polymerase responsible for
the viral replication process has been inferred, although the
rigorous demonstration that such an enzyme is induced and partially coded for by the virus genome has been obtained only for
Qβ replicase. Qβ replicase has been studied in the greatest
detail; consequently, it will be described first--although the
reader is cautioned against concluding that the processes proved
relevant for Qβ replication are necessarily pertinent to the
replication of other RNA viruses. Most of the information gleaned
in vitro about the replication cycle of Qβ replicase has been
obtained through the methodology outlined in previous sections.
Sugiyama et al. [20] comprehensively reviewed transcription,
translation, and replication in many different RNA viruses.

A. Qβ Replicase

Using bacteria infected by the RNA bacteriophage MS2,
Spiegelman and colleagues [21] were able to demonstrate that
cell-free extracts of infected cells possessed an RNA-primed
RNA polymerase activity that was not present in uninfected cells.
The fact that this activity was stimulated by MS2 RNA was con-
firmed in principle by their subsequent observations that a

similar enzyme activity could be obtained from cells after infection by the unrelated Qβ RNA bacteriophage [11]. Purification procedures were developed to free the enzyme extract from
RNA (especially bacteriophage RNA) and to obtain a protein fraction that exhibited a complete requirement for added Qβ RNA for
the RNA polymerase enzyme activity. Other RNA samples, bacterial or viral--even MS2 RNA--were unable to prime this enzyme [22].
Such observations led Spiegelman to propose that bacteriophage
replicases possess template specificity so that they can select
in vivo their own RNA for replication purposes in a milieu of
many other cellular RNA forms.

The purification procedure currently widely adopted for
the purification of Qβ replicase is based on some of Spiegelman's
original procedures, with additional or substitute methods developed in the laboratories of August, Weissmann, and Kamen
[23-26].

Studies of the active principle of the Qβ replicase by
Weissmann and colleagues [27], as well as by Kamen [28], have
indicated that there are four polypeptides involved in the Qβ
replicase enzyme complex--three of them are host-derived and
coded, and the other is a gene product of Qβ RNA. This latter
protein has an enzyme activity that will synthesize poly(G) from
GTP when a poly(C) primer is present. It will also take the Qβ
complementary (minus strand) RNA as template for the purposes
of transcription and make Qβ viral (plus strand) RNA. All four

polypeptides are required for the complete cycle of Qβ viral

RNA replication (i.e., starting template: Qβ viral (plus) RNA;

final product: Qβ viral (plus) RNA).

1. Purification Procedure. The initial Qβ replicase

purification procedure adopted in Spiegelman's laboratories in-

volved protamine sulfate precipitation of nucleic acids from

infected cell extracts, resolution of proteins by DEAE-cellulose

column chromatography and purification by glycerol velocity,

and CsCl equilibrium gradient centrifugation to remove final

traces of DNA-dependent RNA polymerase and trace amounts of

infectious virus particles [23]. August and co-workers intro-

duced an alternative procedure, involving polymer phase separa-

tion of nucleic acids and associated replicase from the infected

cell proteins and release of enzyme by NaCl treatment, as well

as chromatography on hydroxyapatite as a final step in the purif-

ication of the active principle [24]. In either case enzyme

activity was followed through each purification step by the

ability of the protein samples to synthesize RNA on addition of

homologous Qβ viral RNA. Interference in the assay by DNA-

dependent RNA polymerase was reduced by including actinomycin D

(or rifamycin) in the reaction cocktail--neither substantially

affects the Qβ replicase activity. Poly(C)-dependent poly(G)

synthetic activity--requiring only GTP as the nucleoside tri-

phosphate substrate--has also been successfully used to follow

and monitor each purification step. The procedure outlined

below is similar to that recently described by Kamen [25].

Step one: Infected cells, harvested 90 min after addition
of virus to exponentially growing E. coli, can be disrupted by
freeze-thawing three times a 1:1 (w/w) mixture of cells and 0.2 M
sodium chloride, 0.01 M magnesium chloride, 0.5 M 2-mercapto-
ethanol, 0.01 Tris·HCl (pH 7.4) in a dry ice-methanol bath, with
cold water for the thawing stage. Debris is removed from the
suspension by centrifugation for 30 min at 10,000 x g and 4°C.

Step two: Polymer phase separation of the infected cell
supernatant is then performed by adding dextran T500 [1.6%(w/w)
final concentration] and polyethylene glycol (PEG)-6000 [Carbo-
wax 6000, 6.4%(w/w) final concentration]. After dissolving at
4°C for 20 min, the mixture is centrifuged at 10,000 x g for 20
min and 4°C, and the clear, straw-colored polyethylene glycol
supernatant removed. The lower dextran phase is mixed with fresh
buffer and the mixture is adjusted to 2 M NaCl and 5% (w/w)
PEG-6000. After stirring at 4°C, the mixture is resolved by
centrifugation and the top phase, containing replicase activity,
is removed and kept at 4°C. The bottom phase is reextracted
(twice) and the combined PEG phases are dialyzed against 0.1 M
NaCl, 0.5 mM mercaptoethanol, 0.01 M Tris·HCl (pH 7.0) at 4°C
to reduce the NaCl and magnesium chloride concentrations.

Step three: The PEG proteins containing replicase activity
are chromatographed on DEAE-cellulose previously equilibrated
with 0.1 M NaCl, 0.5 mM 2-mercaptoethanol, 0.01 M Tris·HCl (pH
7.0). After application of the protein sample, the column is

washed with the same buffer and then eluted stepwise with buffers
containing 0.12 M NaCl and 0.20 M NaCl. The amounts of DEAE-
cellulose and buffer used depend on the amounts of protein being
processed, the batch, and the source of the DEAE-cellulose. The
majority of the replicase activity is obtained in the 0.20 M
NaCl fractions. The pooled fractions containing replicase
activity are diluted with two volumes of 0.01 M Tris·HCl (pH 7.0),
0.5 mM 2-mercaptoethanol, and applied to a smaller second column,
which is then washed and eluted as before.

Step four: The pooled DEAE-cellulose eluants, after dilution to 0.1 M NaCl final concentration (as before), are chromatographed by loading the solution on a small column of phosphocellulose and eluting stepwise with buffer solutions containing
0.1, 0.2, then 0.3 M NaCl. Most of the replicase activity is obtained in the last eluant, which is then carefully precipitated
with saturated ammonium sulfate buffered to pH 7.0. The purity
of the ammonium sulfate used is important and an "enzyme-grade"
is recommended (e.g., Mann Special enzyme grade). The precipitate is collected by centrifugation and dissolved in a minimal
volume of 0.01 M Tris·HCl (pH 7.0), 0.01 M $MgCl_2$, 0.2 M ammonium sulfate.

Additional steps: Trace amounts of virus or DNA-dependent
RNA polymerase activity can be removed by cesium chloride equilibrium centrifugation or glycerol gradient velocity centrifugation, as described by Pace et al. [23] and Kamen [25].

The enzyme prepared as described above is template-
dependent and stable in 50% (w/v) glycerol at -20°C for months.

2. Properties of the Reaction Products. The availability
of a purified preparation of Qβ replicase allowed a detailed
in vitro analysis of the replication process of Qβ RNA and
postulation of a model of RNA replication reconcilable with
the observations obtained. This work was undertaken in three
laboratories--those of Spiegelman [12], Weissmann [26], and
August [29]--and appears to agree well with in vivo observations of the mechanism of Qβ RNA replication.

It was found, using gel electrophoresis to resolve double-
labeled in vitro reaction products (^{3}H-labeled Qβ template RNA
and [α-^{32}P]UTP to label the product species), that the first
observable product-template complex was partially double-
stranded. The [^{32}P]RNA in these complexes was totally ribonuclease-resistant, whereas the ^{3}H label was mostly ribonuclease sensitive. Subsequently, complexes were obtained in
which although much of the ^{3}H label was still ribonuclease
sensitive, some of the ^{32}P was also ribonuclease sensitive.
After melting, all the [^{32}P]RNA could be annealed to viral (plus)
RNA, indicating that the product species were all complementary
in sequence to the viral genome. It was postulated that the
first product strand was being displaced by subsequent product
strands to give free product single-stranded (i.e., ribonuclease
sensitive) tails, before the template had been completely transcribed. These complexes, containing free product minus strand

tails, moved more slowly upon electrophoresis than did double-
stranded RNA, and with the electrophoretic mobility of known Qβ
multistranded RNA. Shortly after their appearance, and subsequent to further increase in the ribonuclease resistance of the
[³H]RNA, free product single-stranded RNA was obtained having
the same size as Qβ viral RNA but, as shown by hybridization
studies, complementary in sequence to the viral genome. Hence,
it was postulated that the initial three steps in the in vitro
replication process of Qβ RNA resulted in the synthesis of free
Qβ minus-strand RNA.

     Evidence to support the suggestion that free Qβ minus
strands were the direct templates for subsequent synthesis of
product Qβ plus strands by a similar set of reactions was obtained by three different methods. The first demonstrated that
multistranded complexes possessing free plus tails appeared
after the appearance of free Qβ minus strands. Second, the
enzyme, when given purified minus strands, was able to synthesize
plus strands. Third, the direction of minus and plus strand
polymerization was in a 5' to 3' direction, and this result
predicted that before plus-strand initiation there was comple-
tion of a minus-strand template whose 3' terminal was complementary to the 5' of the original template viral plus strand.

     This mechanism of RNA replication posed an interesting
question of how second and subsequent strands were synthesized
on the initial double-stranded template-product complex. Since

double-strandedness was determined on phenol-SDS extracted
nucleic acids, there was considerable interest aroused in this
question when Weissmann and his colleagues showed that the generation of double-strandedness was a function of the extraction
process and that the reaction products in the presence of the
active enzyme (i.e., before, but not after, phenol-SDS extraction) were ribonuclease sensitive [26]. This result suggested,
therefore, that after synthesizing a sequence of complementary
product RNA by hydrogen-bonding, the enzyme separates the product from the template strand and both are then free with respect
to each other. However, upon extraction, because they are in
juxtaposition to each other--due to the replication site--they
can anneal and become double-stranded. Hence, as a result of
this two-stranded freedom, new product strand initiation by a
second or third enzyme molecule is not impaired by interference
from a previous product strand.

The most important demonstration and validification of
studies into in vitro replication processes came from Spiegelman's laboratories, with his demonstration that the product
synthesized by the enzyme was as genetically competent as the
initial viral RNA and represented a net synthesis of new infectious product species [30]. Since Qβ viral RNA is infectious
in E. coli protoplasts, the rigorous demonstration that new
infectious RNA was being synthesized proved, beyond doubt, the
value of such in vitro analyses into the replication mechanisms
of RNA by RNA polymerases.

B. Reovirus Transcriptase

Reovirus is an animal RNA virus that has a segmented
double-stranded genome (10 pieces) [31]. The virion can be converted into a core particle by the action of chymotrypsin; this
core particle possesses [32] an RNA-transcriptase. The transcriptase can also be demonstrated by a brief heat treatment of
the intact virion [33]. The product of the transcription process is repetitive and asymmetric, in that only one strand is
copied. The enzyme responsible for the transcription process
has not been purified. The contributions from many excellent
laboratories on the elucidation of the reovirus transcription
process have been reviewed [34].

C. Rhabdovirus Transcriptases--Particularly Vesicular
Stomatitis Viral Transcriptase

Baltimore and associates demonstrated that the rhabdovirus
vesicular stomatitis (Indiana strain) VSV possesses a virion
transcriptase that synthesizes RNA complementary to the single-
stranded virion RNA genome [35]. In Vitro the process is repetitive and complete [36], although the single-stranded product
RNA species are smaller than the complete 4.4×10^6 dalton
genome [17, 37]. Messenger RNA from VSV-infected cells is also
complementary to the virion genome [38, 39]; thus, the product
of the transcriptase--like that of reovirus--is messenger RNA.
Virion transcriptases of other rhabdoviruses have also been
found: Kern Canyon virus [10], Piry, Chandipura, Cocal, VSV

(New Jersey), Egtved and Systemic Viremia of Carp [unpublished observations]. No virion transcriptase has been found in rabies virus, although this negative result may be due to the in vitro assays used.

D. Myxovirus and Paramyxovirus Transcriptases

The presence of an RNA-dependent RNA polymerase in virions of influenza virus and related myxovirus was demonstrated by Chow and Simpson [40]. The in vitro conditions necessary to obtain enzyme activity and parameters of the product synthesized have been examined [2, 18], and it has been shown that the enzyme only synthesizes RNA complementary to the seven segments of the virion genome. Presumably, this RNA is active as messenger RNA in infected cells—although this fact has yet to be demonstrated. The RNA-containing paramyxoviruses, Newcastle disease virus and Sendai, also possess virion RNA transcriptases [41-43], although the properties of the reaction products have not been studied in as great a detail as those of reovirus, VSV, or influenza viruses.

Many of the questions concerning the role of the virion transcriptases in RNA replication, their association with the viral RNA and other virion proteins, and template specificity have yet to be answered, and provide the goal of those seeking to understand the RNA-dependent RNA polymerases.

ACKNOWLEDGMENT

This study was supported by grant AI 10692 from the National Institute of Allergy and Infectious Diseases.

REFERENCES

[1]   R. H. Symons, Biochim. Biophys. Acta, 155, 609 (1968).

[2]   D. H. L. Bishop, J. F. Obijeski and R. W. Simpson, J. Virol., 8, 66 (1971).

[3]   D. H. L. Bishop, N. R. Pace and S. Spiegelman, Proc. Nat. Acad. Sci. U.S.A., 58, 1790 (1967).

[4]   P. Roy and D. H. L. Bishop, Biochim. Biophys. Acta, 235, 191 (1971).

[5]   A. E. Bresler, Biochim. Biophys. Acta, 61, 29 (1962).

[6]   R. B. Hurlbert and N. B. Furlong, in Methods in Enzymology, Vol. XII (L. Grossman and K. Moldave, ed.), Academic Press, New York, 1967, p. 193.

[7]   I. R. Lehman, M. J. Bessman, E. S. Simms and A. Kornberg, J. Biol. Chem., 233, 163 (1958).

[8]   W. E. Wehrli, D. L. M. Verheyden and J. G. Moffatt, J. Am. Chem. Soc., 87, 2265 (1965).

[9]   I. M. Glynn and J. B. Chappell, Biochem. J., 90, 147 (1964).

[10]   H. G. Aaslestad, H. F. Clark, D. H. L. Bishop and H. Koprowski, J. Virol., 7, 726 (1971).

[11]   I. Haruna and S. Spiegelman, Proc. Nat. Acad. Sci. U.S.A., 54, 579 (1965).

[12]   S. Spiegelman, N. R. Pace, D. R. Mills, R. Levisohn, T. S. Eikhom, M. M. Taylor, R. L. Peterson and D. H. L. Bishop, Cold Spring Harbor Symp. Quant. Biol., 33, 101 (1968).

[13]   D. H. L. Bishop, D. R. Mills and S. Spiegelman, Biochemistry, 7, 3744 (1968).

[14] M. N. Hayashi and M. Hayashi, J. Virol., 9, 207 (1972).

[15] D. H. L. Bishop. J. R. Claybrook and S. Spiegelman,
J. Mol. Biol., 26, 373 (1967).

[16] P. Roy and D. H. L. Bishop, J. Virol., 6, 604 (1970).

[17] D. H. L. Bishop and P. Roy, J. Mol. Biol., 58, 799 (1971).

[18] D. H. L. Bishop, J. F. Obijeski and R. W. Simpson,
J. Virol., 8, 74 (1971).

[19] F. Sanger, G. G. Brownlee and B. G. Barrell, J. Mol. Biol.,
13, 373 (1965).

[20] T. Sugiyama, B. D. Korant and K. K. Lonberg-Holm, Ann.
Rev. Biochem. (in press).

[21] I. Haruna, K. Nozu, Y. Ohtaka and S. Spiegelman, Proc.
Nat. Acad. Sci. U.S.A., 50, 905 (1963).

[22] I. Haruna and S. Spiegelman, Proc. Nat. Acad. Sci. U.S.A.,
54, 1189 (1965).

[23] N. R. Pace, I. Haruna and S. Spiegelman, in Methods in
Enzymology, Vol. XII (L. Grossman and K. Moldave, ed.), Academic
Press, New York, 1968, p. 540.

[24] L. Eoyang and J. T. August, in Methods in Enzymology,
Vol. XII (L. Grossman and K. Moldave, ed.), Academic, New York,
1968, p. 530.

[25] R. Kamen, Biochim. Biophys. Acta, 262, 88 (1972).

[26] C. Weissmann, G. Feix and H. Slor, Cold Spring Harbor
Symp. Quant. Biol., 33, 83 (1968).

[27] M. Kondo, R. Gallerani and C. Weissmann, Nature, 228,
525 (1970).

[28] R. Kamen, Nature, 288, 527 (1970).

[29] J. T. August, A. K. Banerjee, L. Eoyang, M. T. Franze de Ferandez, K. Hori, C. H. Duo, U. Rensing and L. Shapiro, Cold Spring Harbor Symp. Quant. Biol., 33, 73 (1968).

[30] Pace, N. R. and S. Spiegelman, Proc. Nat. Acad. Sci. U.S.A., 55, 1608 (1961).

[31] P. J. Gamatos and I. Tamm, Proc. Nat. Acad. Sci. U.S.A., 49, 707 (1963).

[32] A. J. Shatkin and J. D. Sipe, Proc. Nat. Acad. Sci. U.S.A., 61, 1462 (1969).

[33] J. Borsa and A. F. Graham, Biochem. Biophys. Res. Commun., 33, 895 (1969).

[34] A. J. Shatkin, Bacteriol. Rev., 35, 250 (1971).

[35] D. Baltimore, A. S. Huang and M. Stampfer, Proc. Nat. Acad. Sci. U.S.A., 66, 572 (1970).

[36] D. H. L. Bishop, J. Virol., 7, 486 (1971).

[37] D. H. L. Bishop and P. Roy, J. Mol. Biol., 57, 513 (1971).

[38] J. A. Mudd and D. F. Summers, Virology, 42, 958 (1970).

[39] A. S. Huang, D. Baltimore and M. Stampfer, Virology, 42, 946 (1970).

[40] N. Chow and R. W. Simpson, Proc. Nat. Acad. Sci. U.S.A., 68, 752 (1971).

[41] A. S. Huang, D. Baltimore and M. A. Bratt, J. Virol., 7, 389 (1971).

[42] W. S. Robinson, J. Virol., 8, 81 (1971).

[43] H. O. Stone, A. Portner, D. W. Kingsbury, J. Virol., 8, 174 (1971).

Chapter 2

ISOLATION OF SINGLE-STRANDED VIRAL RIBONUCLEIC ACIDS

T. Sreevalsan

Department of Microbiology
Georgetown University Schools of Medicine and Dentistry
Washington, D.C.

I.   INTRODUCTION

Viral genomes consist of either RNA or DNA.   In recent
years there has been a rapid accumulation of exciting and useful
information concerning the biochemistry of bacterial and animal
RNA-containing viruses.   The current excitement about RNA-containing tumor viruses has lent considerable glamour to the study
of animal riboviruses.   Thus, it is not surprising to find that
techniques for isolation and characterization of viral RNAs have
become increasingly important in virology.

Several methods exist in the literature that are designed
for the isolation of RNAs from cellular or viral structures.
The following account is not an attempt to review the above
methods, but consists of a detailed description of a method we
currently use in our laboratory for the isolation of intact RNAs
from virions or infected tissue-culture cells.   Attempts will
be made to present the rationale behind each step involved in
the procedure.   Our experience with Sindbis virus, an animal
ribovirus, has prompted its use as a model in the discussion.
Sindbis virus is an enveloped ribovirus and is representative
of the majority of animal riboviruses, which are enveloped.
The method to be described here is not new, but consists of

modifications of previously described procedures. Conditions
for the purification and analysis of the viral RNAs by various
methods are not included in the present discussion; they have
been described and discussed elsewhere [1].

II. GENERAL PRINCIPLES

In general, viral RNAs exist in close association with
proteins, either in virions or in infected cells. The main goal
of any procedure for isolation of viral RNA is to remove the
proteins associated with it under conditions that permit minimal
changes in the native state of the RNA molecules. The RNA is
released from the virions or infected cells by exposing them to
agents that denature the viral and cellular proteins. The efficacy of deproteinization depends on the type of agent and the
conditions used. Regardless of the agents or conditions used,
the primary concern of any procedure used for isolation of RNA
is to insure that the RNA molecules remain intact during the
deproteinization and subsequent purification steps. In this
context it is relevant to discuss some of the important factors
affecting the state of RNA molecules during the procedure for
their isolation.

A. Ribonucleases

Single-stranded RNA is highly sensitive to degradation by
ribonucleases. Viral RNA is no exception to this rule. The
viral RNAs, when existing in virions, are protected from the
degradative action of ribonucleases by their close association

with viral proteins; viral RNAs freed of proteins become extremely susceptible to enzymatic hydrolysis by ribonucleases.
This should be remembered at all times, especially during the
isolation of viral RNA.

When one considers the fact that ribonucleases exist in
various tissues, cell fragments, and subcellular particles, it
is obvious that they create a problem. Ribonuclease is one of
the enzymes contained in lysosomes of cells and tissues. Lyso-
somal ribonuclease is usually contained in preparations of
virions or cytoplasmic extracts obtained from infected cells.
Ribonucleases appear to adhere to virions and are not easily
removed from them during purification procedures. The type and
the amount of ribonuclease present in cells varies with the type
of cells or tissues used. Generally, two types of ribonucleases
can be recognized, based on their pH optima, in animal tissues;
one RNase acts optimally between pH values of 7 and 8 and the
other between 4.5 and 6.0 [2]. Both types are heat stable.
Penman et al.[3] described the presence of a soluble RNase (pH
optimum 8.0) and a particulate RNase (pH optimum 5.0) in cytoplasmic extracts of HeLa cells. The relative stability of
ribonucleases at higher temperatures (above 37°C), their pH optima, and their relative resistance against protein denaturing
agents are factors that contribute to the difficulty of estab-
lishing conditions wherein the degradative action of ribonucleases
can be minimized during the isolation of RNAs. The problem of

ribonucleases during the isolation of RNA is further complicated by the fact that ribonucleases show an unusual affinity for
glass surfaces [4]. All of these considerations about ribonucleases should be kept in mind when one attempts to isolate intact
RNA from any biological source. Despite the use of various
inhibitors or inactivating agents for ribonucleases, there is
no "universal" procedure available to guarantee the absence of
degradative action by contaminating ribonucleases on RNA during
its isolation, purification, handling, or storage. Various
nuclease inhibitors like polyvinyl sulfate, bentonite, and
diethylpyrocarbonate are used to inhibit any enzymatic degradation of the RNA. Also, enxymatic hydrolysis of the RNA can be
minimized by manipulation of the conditions used for isolation
of RNA, namely, pH, ionic strength, temperature, and metal ions.
These considerations are amply dealt with elsewhere [1,4].

B. pH and Ionic Strength

RNA is degraded at pH values above 10.0, since its phosphodiester bonds are hydrolyzed. Similarly, at pH values below
3.0, RNA slowly undergoes depurination. Thus, it is important
that extreme pH conditions be avoided during the isolation and
purification of viral RNAs. The secondary structure of RNA is
dependent on the ionic strength of the solvent in which the
nucleic acid is suspended. RNA is precipitated from solutions
containing salt at 1.0 M or higher concentrations. Similarly,
in low-salt environments (0.1 mM to 1 mM), RNA molecules possess

little or no secondary structure. Loss of secondary structure
leads to increased susceptibility of RNA molecules to the
action of ribonuclease. Thus, use of very high or very low ionic
strength buffers in the isolation procedures are not recommended.
Also, the presence or absence of cations in the buffer used in-
fluences the state of RNA molecules. Magnesium ions at low
concentration (0.1 - 1 mM) stabilize polynucleotide chains
through base-to-base interactions. However, at higher concentrations (10 - 100 mM) Mg ions cause aggregation of RNA molecules.

C. Temperature

The denaturation of proteins and DNA occur faster than that
of RNA at higher temperatures. Thus, the isolation of viral RNA
free of proteins and DNA from infected cells or tissues is
usually more efficient at higher temperatures (60°C) than at
ambient temperatures. However, the use of higher temperatures
for isolation of RNA introduces a potential for the degradation
of viral RNA, since RNA molecules undergo thermal hydrolysis
at such temperatures. Also, enzymatic degradation of RNA by
nucleases is more likely to occur at higher temperatures (37°C
or above). The combined use of phenol and sodium dodecyl sulfate (SDS) at 25°C appears to partially solve this dilemma.
Experience in our laboratory has indicated that extraction procedures for RNA that utilize high temperature (60°C) usually
lead to the partial degradation of high molecular weight viral
or cellular RNAs. Such nonspecific degradation can be avoided

by use of ambient temperature during the isolation of RNAs.
Thus, elevated temperatures (37°C or higher) are not recommended
during any step in the procedure for isolation of RNAs.

D. Deproteinizing Agents

The major step involved in the isolation of RNA from
virions or infected cells is the removal of associated proteins, usually by the use of agents capable of denaturing the
proteins. The efficiency of any isolation procedure depends on
the capacity of the above agents to liberate the nucleic acid.
Several agents are used for this purpose. Phenol is the most
common denaturing agent and is preferred by several investigators. Methods that use phenol appear consistently to yield RNA
preparations with the minimal amount of proteins. Such methods
have been used in several instances to isolate infectious viral
RNA from many RNA viruses. The exact mechanism of the denaturing action of phenol on proteins is poorly understood. Ralph
and Bergquist [4] suggested that phenol may denature and precipitate globular proteins by destroying their internal organi-
zation, which results from the interactions among the various
aliphatic, aromatic, and heterocyclic groups on the molecules.

Besides phenol, other agents may be used to denature viral
or cellular proteins. Among the many known detergents SDS is
the most widely used. SDS solubilizes the proteins and liber-
ates the nucleic acids. The use of SDS to isolate RNA is ideal,
since it also acts as a potent inhibitor of ribonucleases.

However, when SDS is the sole agent used for isolation of the
RNA, the protein remains in solution along with the liberated
RNA. Therefore, solubilized proteins have to be removed by
methods that separate proteins from RNA. One of the approaches
toward the above purpose is the use of SDS in conjunction with
phenol. SDS acts as the solubilizing agent, while phenol dena-
tures and precipitates the solubilized proteins, leaving the
nucleic acid free in solution. The only disadvantage to the
use of SDS is its property of precipitating at low temperatures
and/or in the presence of potassium ions. Detergents like
Sarkosyl, tri-isopropyl naphthalene, have been used successfully in many instances. However, SDS appears to be the detergent of choice, especially when used in conjunction with phenol.

III.   ISOLATION OF VIRAL RNAs FROM SINDBIS VIRIONS
OR TISSUE CULTURE CELLS INFECTED WITH SINDBIS VIRUS

A.   Principle of the Procedure

In principle, the method consists of a modification of the
phenol-SDS method. Pronase is used to degrade the viral or
cellular proteins. Pronase is a mixture of several endo and exo-
peptidases that is produced by a strain of Streptomyces griseus
K-1 [5]. It has been used successfully by others to isolate
high molecular weight DNA from various sources [4,6,7]. The
specific advantage in the use of Pronase during isolation procedures lies in the fact that proteins, including ribonucleases,
are degraded during incubation of the materials with the enzyme.

As pointed out earlier, the major difficulty encountered during
the isolation of RNA is the potential release of ribonucleases
during the proteinization step. Prior incubation of virions or
infected tissue culture cells with ribonuclease-free Pronase will
degrade the released ribonucleases. Such enzymatically degraded
viral or cellular materials can be treated with phenol and SDS.
The above procedures ensure minimal degradation of the viral RNA
by nucleases during the isolation procedure. Our experience
has indicated that the enzyme-detergent-phenol method has consistently yielded undegraded RNAs from various sources, including virus-infected tissues and cells. Use of the SDS-phenol
method under identical conditions yielded variable amounts of
intact RNAs. One of the disadvantages of the enzyme-SDS-phenol
method is the failure to remove cellular DNA from the viral RNA
when samples originate from infected cells or tissues. However,
this is not a serious drawback since the DNA present in the deproteinized samples can be easily hydrolyzed enzymatically by
DNase. Ribonuclease-free DNase is essential for the above
step; such preparations of DNase are available commercially.
Removal of DNase from the final preparation of the RNA depends
on its subsequent use. Thus, if the RNA is to be used for
analysis by sucrose density gradients or electrophoresis, then
the DNase can be solubilized by the addition of SDS (to 1%)
before use. However, if the RNA is used for hybridization
studies, the DNase should be removed by extraction with phenol.

B. Materials Needed

All glassware used for the isolation, purification, or storage of RNA are sterilized whenever possible. This step ensures the inactivation of ribonucleases that are usually found on the surface of glassware. Sterilization is usually accomplished by dry heat at 162°C overnight or autoclaving for 20 min at 20 lb/in^2 pressure. Similarly, wherever possible, the solutions or reagents used in the procedure are made up in sterilized and deionized water.

1. Equipment

a. Glass Centrifuge Tubes: 15, 30, or 150 ml Capacity.

Corex glass tubes (Corning) are recommended, since they are made of aluminosilicate glass and can be used for centrifugation at high speeds.

b. Disposable Capillary or Pasteur Pipettes.

These pipettes are convenient for withdrawing the phenol below the aqueous phase, and also can be used to stir the mixture during the extraction step with phenol. They should possess tips that are long and large in diameter.

c. Water Bath.

Any ordinary water bath deep enough to hold samples is appropriate.

d. Shaker.

A Vortex mixer is convenient for shaking samples intermittently and vigorously during the extraction procedure.

e. Centrifuge. For the extraction procedures, we use an angle-head rotor (SS-1), Sorvall, Norwalk, Connecticut. Any high-speed centrifuge (International, Sorvall, or Beckman J-21) can be used. Refrigeration during centrifugation is optional in the extraction procedure. However, during recovery of the alcohol precipitate, RNA is centrifuged at 4°C.

2. Chemicals and Enzymes

a. Pronase. We use commercially available enzyme preparations without further purification. Pronase is purchased from Calbiochem, Los Angeles, Cal. (Catalog #53702, B grade) and is processed according to the method of Stern [8] to remove the ribonuclease present. Stock solutions of the enzyme are prepared as follows. Pronase is dissolved in sterile and deionized water to 2 mg/ml. The pH of the solution is adjusted to 5.0 with 1 N HCl. The solution is heated to 80°C for 20 min, cooled to 37°C, and held at that temperature for 2 hr. The above steps are intended for hydrolysis of any contaminating ribonuclease present in the preparation. The solution is then made 0.01 M with respect to Tris (pH 7.0) and adjusted to pH 7.0 with 1 N NaOH. Solid NaCl is added, to a final concentration of 1.0 M. The solution is then distributed (1-ml aliquots) into sterile glass tubes and stored at -20°C. Pronase preparations processed by the above method can be stored at -20°C for at least six months without appreciable loss of proteolytic activity.

b. Phenol. Commercially available phenol is satisfactory if redistilled. The distilled phenol is saturated with
an aqueous solution of 1 mM EDTA, pH 7.4, and aliquots are
stored frozen at -20°C in amber-colored bottles. Storage of
phenol at room temperature is not recommended, since a brown
color develops. Discolored batches of phenol are not used.
Chelating agents such as 8-hydroxyquinoline (0.1%) can be added
to the buffer-saturated phenol before storage, as they prevent
the occurrence of metal ion-catalyzed oxidation. Just before
use, the frozen phenol is thawed by immersion of the bottles in
warm water; the melted phenol can be stored at 4°C for only
one week.

c. Sodium Dodecyl Sulfate (SDS). Any high grade
(recrystallized) preparation can be used. Stock solutions (10%)
of the detergent are made up in sterile water and refrigerated.
The solution is warmed before use since SDS precipitates at low
temperatures.

d. Deoxyribonuclease (RNase free). Preparations of
the enzyme are purchased from Worthington, Freehold, N.J. Each
batch of DNase is assayed for ribonuclease before it is used.

e. Yeast RNA. A commercial sample of RNA is purified by extraction with phenol three times, followed by precipitation with alcohol. This step removes the contaminating ribonuclease present in the commercial preparations. The yeast RNA
is used as a carrier for precipitation of extremely low concentrations of RNA with alcohol.

f. __Ethanol__. 95%, without any additives.

3. __Solutions and Buffers__

a. __Phosphate-Buffered Saline (PBS)__. This is prepared by the method of Dulbecco and Vogt [9]. Antibiotics and phenol red are omitted from the solution.

b. __Tris-EDTA-Saline (TES)__. The solution contains 0.1 M NaCl, 1 mM ethylenediaminetetraacetate (EDTA), and 10 mM Tris(hydroxymethyl)aminomethane (Tris), at pH 7.4. The solution is used for suspension of nucleic acids and also for the preparation of sucrose solutions used in the analysis of RNA samples.

c. __EDTA Solution__. This solution consists of 0.01 M EDTA-Na salt in sterile distilled water. The pH of the solution is adjusted to 7.4 with 1 N HCl. The solution is used to suspend the purified virus or infected cells at the start of the procedure for isolation of RNA.

d. __Tris-MgCl$_2$-Saline__. The above solution contains 0.05 M Tris, pH 7.4, 0.1 M NaCl, and 1 mM MgCl$_2$. It is used to dissolve the nucleic acid before incubation with ribonuclease-free DNase.

C. __Sample Preparation for Viral RNA Isolation__

The RNA content of all but a few viruses represents only a minor proportion of the total viral constituents. Therefore, attempts to isolate, purify, and characterize viral RNAs from mature virions or infected cells necessitates labeling the RNA with a precursor of RNA. The method and conditions used for radiolabeling viral RNAs depend on the type of virus and host cell system used. The reader is referred to the excellent review on this subject by Henry [10]. In our laboratory, labeled Sindbis virus or virus-infected cells are prepared as follows:

The multiplication of Sindbis virus is insensitive to actinomycin D, a drug that inhibits the synthesis of cellular RNA. Therefore, actinomycin D can be used in experiments involving the specific radiolabeling of viral RNAs. Monolayers of primary chick-embryo cells are incubated at 37°C for 1 hr with Eagle's medium containing 5 µg/ml of actinomycin D (Merck, Sharp and Dohme). The monolayers are then washed with PBS and infected with Sindbis virus at a multiplicity of 10 plaque-forming units (PFU) per cell (adsorption at 25°C for 30 min). The cultures are washed three times with PBS. Prewarmed Eagle's medium containing [^{3}H]- or [^{14}C]uridine and 2% dialyzed calf serum is added to the cultures, which are incubated at 37°C. In experiments involving the preparation of labeled virus, the cultures are incubated for 24 hr, then the medium is collected and used for purification of the virus by methods described elsewhere [11]. In experiments dealing with virus-specific RNAs synthesized in the infected cells, the cultures are harvested at appropriate intervals after infection. Monolayers of cells are washed with cold PBS. The cells are collected by scraping the monolayers into cold PBS with a sterile rubber

policeman. The cells are pelleted by centrifugation at 800 x g
for 10 min at 4°C in a refrigerated centrifuge. The pellet of
cells is used immediately for the isolation of viral RNAs.

The methods described above are applicable only to Sindbis
virus or other riboviruses with similar properties. The conditions and procedures for the preparation of virus particles and
infected cells vary with the kind of ribovirus under study.
These should be established for the type of virus under investigation before attempts are made to isolate the RNA.

D. Isolation of Viral RNA

RNA is isolated from radiolabeled virus particles or infected cells in the following manner: samples containing the
virus particles are suspended at a concentration of 1 mg/ml in
the solution of 0.01 M EDTA, pH 7.4. When infected cells are
the starting material, the cell pellet should be suspended in
the EDTA solution at a concentration of 2-4 x 10^7 cells per ml.
The sample is shaken briefly on a Vortex mixer and immediately
the stock solution of Pronase is added to the above suspension
to give a final concentration of 500 µg/ml. Then, the sample
is incubated at 35°C for 15 min, during which time enzymatic
degradation of the viral or cellular proteins takes place. At
the end of the incubation period, a stock solution of SDS is added
to a final concentration of 0.5%. The sample is incubated for
an additional 15 min at 35°C for the complete degradation and
dissolution of the proteins by the combined action of Pronase

and SDS. An equal volume of phenol (saturated with 0.01 M
EDTA, pH 7.4) is added to the sample and mixed vigorously by
intermittent mixing on a Vortex mixer. The mixture is incubated
at 25°C for 3 min. Aliquots may then be removed for the determination of total radioactivity in the samples. The mixture is
centrifuged at 10,000 x g for 1 min. The centrifugation step
aids in breaking the emulsion between phenol and the aqueous
solvent. The clear phenol layer (bottom layer) is removed
completely by the use of a capillary pipette, leaving the
aqueous layer, the interphase, and any pellet in the centrifuge
tube. The phenol layer can be discarded since it contains no
significant amount of RNA. The solution remaining in the centrifuge tube is mixed vigorously on a Vortex mixer, and an equal
volume of fresh phenol is added. The mixture is shaken vigorously as before and extraction is continued at 25°C for 3 min.
After centrifugation and removal of the phenol, the sample is
resuspended on a Vortex mixer and again extracted with fresh
phenol.

At the end of the third extraction, little or no pellet is
visible. However, a considerable amount of interphase, consisting chiefly of DNA and some RNA, still remains, Most of the
RNA remains in the aqueous layer. The aqueous layer, as well
as the interphase, is transferred carefully into a fresh centrifuge tube, capable of holding at least three times the total
volume of the sample. Two volumes of ethanol cooled to -20°C

are added and mixed by gentle stirring or by sucking and expel-
ling the mixture with a capillary pipette. The mixture is
then incubated at -20°C for at least 2 hr for complete precipitation of the nucleic acids. If the concentration of the starting material is too low to form a visible precipitate, we usually
add 100 μg of unlabeled yeast RNA to the sample before the addition of alcohol. The added RNA serves a dual function. The
second function is to prevent the possible denaturation of RNA
induced by alcohol when low concentrations of RNA are exposed
to alcohol.

The precipitate formed after the addition of alcohol is
recovered by centrifugation of the sample (in an angle-head
rotor) at 10,000 x g for 15 min at 4°C. The ethanol is poured
off carefully so that the pellet is not disturbed. The tube
containing the pellet and the small amount of alcohol is centrifuged again as before for 5 min. The alcohol can now be
removed with a long capillary pipette. The centrifuge tube is
slanted without disturbing the pellet and the remaining liquid
can be drained by the use of a clean paper towel (we usually
autoclave the paper towels before use). The pellet is dissolved rapidly in Tris-NaCl-MgCl$_2$ buffer in a volume 5 to 10 times
the original volume of the cells. At this stage the solution
may be viscous and the precipitate may not be in complete solution due to cellular DNA present in the sample. Enough stock
DNase solution is added to make a final concentration of 100 μg/ml.

The sample is incubated at 37°C for 30 min to degrade the DNA. At the end of the incubation, the viscosity of the sample will be considerably reduced due to hydrolysis of DNA. The DNase present in the sample is removed by extracting three times with phenol at 25°C, according to the steps described in Section III,D. At this stage, little or no interphase can be seen in the mixture because of the removal of DNA. The RNA contained in the aqueous phase is precipitated with alcohol and dissolved in TES buffer. Contamination with DNA does not exist when RNA is isolated from virus particles only. Therefore, in such instances the steps involving incubation with DNase and re-extraction with phenol are unnecessary.

Viral RNA, recovered as a precipitate after the above steps, contains little or no DNA. However, the solution of RNA still contains traces of phenol, since a single step of precipitation with alcohol fails to completely eliminate phenol. Traces of phenol present in the sample should be removed when the RNA is used for determination of infectivity or absorbance. Phenol is toxic to the tissue culture cells used for assays of infectivity. Also, phenol interferes with determination of absorbance of RNA since it possesses absorbancy at wavelengths at which nucleic acids also absorb. However, if the RNA is used directly for analysis on sucrose density gradients, contaminating phenol does not pose a problem. Contaminating phenol is eliminated from the samples by reprecipitation of RNA at -20°C by the addition

of 2 volumes of ethanol. The pellet of RNA obtained after reprecipitation is recovered by centrifugation. The traces of
alcohol adhering to the pellet are removed by a short exposure
of the RNA to a high vacuum. This step is valuable--especially
if the RNA is used for electrophoretic analysis--since amounts
of alcohol present in the sample tend to shrink the polyacrylamide gel at the point of application of the sample. The RNA,
once freed of traces of alcohol, can be dissolved in appropriate
volumes of TES buffer and used for further analysis.

E. Isolation and Analysis of RNA from Virus Particles

The enzyme-SDS-phenol method can be used successfully for
isolation of RNA from virions or infected cells. However,
RNA from virus particles can be quickly characterized by the
following method, which is simple and less tedious than that
described above. The method to be described serves mainly as
an analytical, rather than a preparative, procedure.

A sample of the radiolabeled and purified virus is incubated with 0.01 M EDTA, Pronase, and SDS as described in Section D.
For Sindbis virus we use directly an aliquot of the tissue
culture medium obtained from radiolabeled and infected cells
for analyzing the viral RNA. Such a procedure is possible
since Sindbis virus is released into the medium subsequent to
its maturation at the cellular surface. At the end of the 30-
min incubation period at 37°C, the sample is layered directly
onto the top of a gradient of sucrose in TES buffer containing
0.5% SDS. The gradient is spun at 25°C in a swinging-bucket

rotor in any of the commercially available centrifuges that
allow speeds up to 50,000 rpm. The temperature at which the
gradient is centrifuged is critical, since SDS present in the
sample and the gradient will precipitate at low temperatures.
The type of sucrose density gradient, the rotor, and the time
of centrifugation are dependent on the density of the RNA
used, as well as the degree of resolution sought. A detailed
account of the parameters influencing the sedimentation of RNA
on sucrose density gradients is available [1]. (For the
analysis of Sindbis viral RNA, we use a 32-ml gradient of 5
to 20% sucrose in TES buffer with 0.5% SDS. The gradient is
spun at 25°C in a SW 25 Beckman rotor at 62,000 x g for 14 hr.)
Fractions are collected from the gradient after centrifugation,
and the radioactivity present in the samples is estimated [11].
The results of a typical experiment are presented in Figure 1.
The distribution of radioactivity in the gradient demonstrates
a sharp and clean peak, suggesting little or no degradation of
the viral RNA. Such a conclusion is further substantiated
from the electrophoretic behavior on polyacrylamide gels of
the RNA obtained after analysis of the radiolabeled material
contained in the peak fraction from the scurose gradient.
Analysis of the virion RNA by this method compared quite favor-
ably with that obtained by the enzyme-SDS-phenol method (compare results in Figure 2a. The only drawback to the above
method is its limited applicability as a preparative procedure

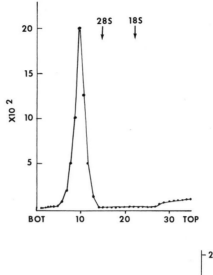

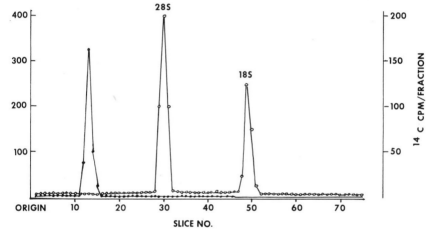

Fig. 1.   Analysis of RNA from radiolabeled Sindbis virus.
[³H]Uridine-labeled virus was prepared by infection of chick-
embryo cultures with Sindbis virus in the presence of actinomycin D (Section III, C).   An aliquot of the medium was removed
from the infected cultures after 24 hr of incubation.   The
medium was centrifuged at 10,000 x g for 10 min to remove
cellular debris.   The clarified medium was incubated with Pronase
and SDS (see Section III, E), incubated at 37°C for 15 min, and
immediately layered on a 32-ml gradient of sucrose (5 to 20%)
made in TES buffer containing 0.5% SDS.   The gradients were

for isolation of viral RNA. The SDS contained in the sucrose
density gradient interferes with attempts to recover or concentrate the viral RNA present in the appropriate fractions by
precipitation with alcohol, since SDS precipitates at low temperatures. Additionally, only limited volumes of samples can be
analyzed on sucrose density gradients without sacrificing the
analytical resolution of RNA. However, the ease of the procedure and the integrity of the viral RNA obtained make the above
technique a simple and rapid one for the analysis of RNA present
in virions.

Fig. 1-continued

centrifuged at 24,000 rpm in a SW 25 rotor at 25°C for 14 hr.
Fractions were collected and aliquots were used for the determination of acid-insoluble radioactivity (insert). The arrows
indicate the positions where the 28S and 18S ribosomal RNAs from
chick-embryo cells sediment in sucrose density gradients under
the above conditions.

The results shown in the bottom part of Fig. 1 represent an
electropherogram of the labeled viral RNA contained in Fraction
10 of the sucrose density gradient. The sucrose density gradient
fraction 10 was incubated with 2 volumes of alcohol and 50 μg of
yeast carrier RNA at -20°C (Section III,D). The precipitated
RNA was recovered and analyzed on a 2.2% polyacrylamide-agarose
gel, according to the method of Peacock and Dingman [15]. [^{14}C]-
Uridine-labeled 28S and 18S ribosomal RNAs from chick-embryo
cells were coelectrophoresed with the viral RNA.

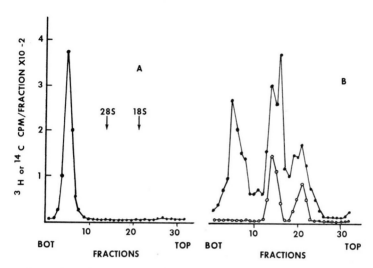

Fig. 2. Sucrose density gradient analysis of labeled RNA
isolated from (A) Sindbis virus and (B) infected chick-embryo cells
by the enzyme-SDS-phenol method. [^{3}H]Uridine-labeled virus or
infected cells were obtained as described in Section III,C.
The isolated viral RNAs were analyzed on 16-ml sucrose density
gradients (5 to 20%) made in TES buffer. Centrifugation was at
4°C for 16 hr at 22,000 rpm. Fractions were collected and
assayed for acid-insoluble radioactivity. Where indicated,
[^{14}C]uridine-labeled 28S and 18S ribosomal RNAs were cocentri-
fuged with the viral RNAs on the sucrose density gradient.

The results presented in Fig. 2b also show the analysis of
Sindbis viral RNAs obtained from infected cells in the fractions
obtained after centrifugation of the viral RNAs on sucrose

gradients. It can be seen that in infected cells there exist other types of viral RNAs besides the RNA found in virions. These RNAs do not represent degradation products. It has been established by us, as well as by several others, that infection of cells with Sindbis virus induces the synthesis of at least two species of single-stranded RNAs, possessing sedimentation rates of 26S and 16S, besides the replicative intermediate and replicative forms of viral RNAs [11 - 14]. Thus, the results presented here (Figs. 1 and 2) indicate that undegraded viral RNAs can be generated from infected cells or virus particles by use of the procedures described here. Additionally, the methods we have described yield infectious RNA from the virus particles. Infectivity of viral RNA molecules is a good indicator of molecular integrity, since it has been shown in several instances that a single break in an RNA molecule destroys its infectivity.

IV. SOME COMMENTS AND USEFUL TIPS

So far, we have dealt with the important parameters influencing the isolation of viral RNA. Minor details often play a major role in the success of any procedure, and we shall discuss them in this section. Some of them may appear unworthy of mention. However, the author's past experience has indicated that close attention to the minor details often pays off in the final product.

A. Presence of Ribonuclease on the Skin

Despite the use of many rigid conditions to minimize

contamination with ribonucleases, one of the possible reasons
for failure to isolate intact viral RNA is careless technique.
The surface of human skin carries traces of ribonuclease that
can be easily transferred into RNA samples during isolation
and handling. Thus, extreme precaution should be taken to avoid
direct contact with the interior of glassware or other materials
used in the procedure. This applies especially in situations
where the tips of pipettes, the top interior part of centrifuge
tubes, or other similar materials are accidentally touched by
bare fingers or skin. The best way to prevent such accidents
is to use sterile gloves and to discard glassware that has been
suspected of being in contact with the skin. The use of gloves
also helps to prevent accidental transfer of phenol to one's
skin. Additionally, propipettes or capillary pipettes should
be used when solutions of phenol are transferred. Pipetting
phenol by oral suction is potentially dangerous, and should be
avoided as much as possible.

B. Storage and Handling of Samples

Ideally, samples used for the isolation of viral RNA
should be processed as fast as possible. This applies especially in cases where the source of the sample for isolation of
viral RNA is infected cells or subcellular structures. The endogenous nucleases that can be released during the lysis of
cells will rapidly degrade the viral RNA contained therein.
So, as a general rule, once the cells are lysed it is necessary

to carry out, without interruption, all the procedures up to
and including the initial precipitation of RNA with alcohol
(Section III,D). In experiments where samples are obtained at
various time intervals, the procedure used is as follows:
Monolayers are washed with cold PBS and stored at 4°C in a
bucket containing crushed ice. The cells are harvested from
the monolayers only after the last sample is taken. The possibility of breakdown of viral RNA is highest at the initial step,
the suspension of the cells or virus in 0.01 M EDTA, pH 7.4.
Therefore, rapid mixture of the suspension, followed by immediate addition of Pronase and incubation at 35°C, are highly recommended.

C. Separation of Phenol and Aqueous Phases

Intermittent, yet vigorous, mixing of phenol with the
aqueous phase is necessary for efficient deproteinization of
samples; we place centrifuge tubes on a Vortex mixer for 30 sec.
However, care should be taken to avoid spillage or spitting of
the mixture from the centrifuge tube since, besides loss of the
sample, the phenol present in the mixture can spill on the
face and eyes. These hazards can be overcome by the use of
stoppered centrifuge tubes capable of holding the sample conveniently. Chilling the emulsion containing phenol and the
sample for a minute in a dry ice-alcohol bath just before
centrifugation will hasten breakage of the emulsion and the subsequent separation of the two phases.

D. Interphase

The interphase that accumulates during the extraction procedure should be vigorously mixed before the addition of fresh
phenol. The interphase consists of a mixture of DNA, RNA, and
protein; it should not be discarded since doing so will decrease
the final yield of RNA. A very heavy interphase results usually
from an inadequate amount of EDTA solution being used for the
lysis of the infected cells (Section III,D). Additional
amounts of SDS can be added at the second cycle of the deproteinization step. The pellet that is sometimes found in the
clear layer of phenol subsequent to breakage of the emulsion
by centrifugation (Section III,D) consists of an artificial
mixture of DNA, RNA, protein, and partially lysed cells. Such
pellets should not be discarded during the removal of the
phenol, but should be mixed with the remaining interphase and
aqueous phase before the second cycle of extraction with phenol
is initiated.

E. Precipitation with Ethanol

Storage of RNA in the form of a precipitate in alcohol at
-70°C appears to minimize the nonspecific degradation that
usually accompanies prolonged storage of RNA. However, this
procedure is not recommended when the concentration of RNA is
extremely low (below 10-20 µg) since denaturation of the RNA by
alcohol can occur. This problem can be overcome when one deals
with radiolabeled RNA, since unlabeled yeast RNA can be added

to raise the concentration of RNA in samples. Similarily, the
final concentration of alcohol used in the step involving precipitation of the RNA should not exceed 66%, since at higher
concentrations of alcohol, polysaccharides present in the deproteinized samples precipitate out. Traces of alcohol can be
removed from the pellet of RNA by plugging the open end of the
tube with a sterile cotton swab and storing the tube in an inverted position at -20°C overnight. However, care must be taken
to ensure that the pellet of RNA is tightly packed.

F. Presence of Protein in RNA Preparations

Inefficient deproteinization leads to heavy contamination
with proteins in the final preparation of RNA. The degree of
contamination of RNA with protein can be assessed roughly by
measurement of the absorbance of the final preparation. Absorbance of the samples (diluted with water to achieve concentrations
that will yield reliable values) is measured at 230, 260, and 280
nm. RNA samples contaminated with 0.1% or less of proteins will
give a 260 to 230-nm ratio higher than 2.3, while the 260 to
280-nm ratio will be 2.0 or more. An absorbance of 24.0 units
at 260 nm in a 10-mm light path usually represents about 1 mg/ml
of RNA. The presence of unusually large amounts of protein in
RNA samples will interfere if such RNA is used for analysis by
electrophoresis on polyacrylamide gels. The contaminating proteins tend to alter the relative electrophoretic mobility of
the RNA on acrylamide gels. Therefore, samples of RNA intended
for electrophoresis should be checked for contamination with
proteins before use.

ACKNOWLEDGMENT

This work was supported by a Public Health Service Research Grant AI-09355-02 from the National Institute of Health.

REFERENCES

[1] K. Scherer, in Fundamental Techniques in Virology (K. Habel and N. P. Salzmann, eds.), Academic Press, New York, 1969, p. 413.

[2] C. B. Anfinsen and F. H. White, in The Enzymes (P. D. Boyer, H. Henry, and K. Mayerback, eds.), Vol. 5, Academic Press, New York, 1961, p. 95.

[3] S. Penman, H. Greenberg, and M. Willems, in Fundamental Techniques in Virology (K. Habel and N. P. Salzman, eds.), Academic Press, New York, 1969, p. 49.

[4] R. K. Ralph and P. L. Bergquist, in Methods in Virology (K. Maramorosch and H. Koprowski, eds.), Vol. II, Academic Press, New York, 1967, p. 463.

[5] Y. Narahashi, in Methods in Enzymology (S. P. Colowick and N. O. Haplan, eds.), Vol. XIX, Academic Press, New York, 1970, p. 651.

[6] W. Joklik, J. Mol. Biol., 6, 26 (1962).

[7] C. J. Pfau and J. F. McCrea, Nature, 194, 894 (1962).

[8] H. Stern, in Methods in Enzymology (K. Moldave and L. Grossman, eds.), Vol. XIIB, Academic Press, New York, 1968, p. 100.

[9]   R. Dulbecco and M. Vogt, J. Exp. Med., 99, 183 (1954).

[10] C. Henry, in Methods in Virology (K. Maramorosch and H.
Koprowski, eds.), Vol. II, Academic Press, New York, 1967, p. 427.

[11] T. Sreevalsan, J. Virology, 6, 438, (1970).

[12] B. W. Burge and E. R. Pfefferkorn, J. Mol. Biol., 35, 193
(1968).

[13] F. H. Yin and R. Z. Lockart, J., J. Virology, 2, 728 (1968).

[14] R. G. Levine and R. M. Friedman, J. Virology, 7, 504 (1971).

[15] A. C. Peacock and C. W. Dingman, Biochemistry, 1, 668 (1968).

Chapter 3

RNA POLYMERASE IN HIGHER PLANTS

Rusty J. Mans

Department of Biochemistry
University of Florida
Gainesville, Florida

I. INTRODUCTION

This discussion of RNA polymerases in plants is limited to

93

a description of methods used in this laboratory for the detection and isolation of polymerization activities in corn seedlings. No review of methodology is intended, but rather this
chapter is intended to be a presentation of what works for us,
in enough detail to serve as a model for the reader's own research
pursuits. No effort has been made to eliminate the author's
prejudices or quirks; therefore, the reader is cautioned not to
consider this presentation as that of an authority in the field,
but rather as that of a practioner of the enzymological approach
to the resolution of biological problems. Even after resorting
to the literature, I am often at a loss as to how to initiate
an experimental approach. Having read this chapter on RNA polymerases in plants, I hope the reader will be titillated to initiate his own enzymological sojourn.

There are three prerequisites for embarking upon purification of an enzyme. First, you must have a well-defined experimental objective that cannot be attained except by enzyme purification. Second, you must have an easy, quick, specific, and
inexpensive assay for the enzymatic activity of interest. An
increase in the rate of product accumulation must be a function
of an increase in enzymatic protein. An assay for protein equal
in sensitivity to that for activity is required. A method to
determine directly the relative purity of a protein preparation
(polyacrylamide gel electrophoresis) in the course of purification is advantageous. The third prerequisite is an adequate and

appropriate source of starting material to progress through the

intermediate steps during purification and to insure enzyme iso-

lation from a tissue that serves your experimental objective.

Ancillary considerations include: adequate equipment not only

for isolation and purification (homogenizers and centrifuges)

but for handling raw materials, and for cold storage of prep-

arations in the intermediate and final stages of purification.

Appropriate instrumentation for assays must be conveniently

available. I consider an ultraviolet spectrophotometer for

assays, column monitoring, and gel scanning, and a liquid scintilla-

tion spectrometer for isotopic assays essential tools for poly-

merase purifications.

II. RATIONALE AND HISTORY

Understanding of the regulation of biosynthetic pathways

of cells continues to be a focal point of research in the biol-

ogical sciences. The mode of genetic regulation of transcription

and the consequent control of protein synthesis would seem to be

basic to understanding other regulatory processes within the

cell. We now have the technical competence to dissect and iso-

late components of the nucleic acid synthetic machinery from

the cell, place them in a controlled environment, and investi-

gate the mechanism(s) of genetic regulatory control that act

upon them. Requisite to our understanding the regulation of

multicomponent processes is an intimate knowledge of the compon-

ents and their interactions. Cell-free systems derived from

mammalian, microbial, and higher plant tissues have contributed
immeasurably to our current concepts of protein, DNA, and RNA
synthesis, and of interactions among various components regulating and directing these biosynthetic processes. Depending upon
the level of sophistication, and how faithfully the in vivo
situation is reconstructed in such model systems, results and
inferences from in vitro studies can be extended to intact organisms.

Ten years ago it seemed reasonable to ask: "Could it be
that external stimuli, such as light, or internal regulators,
such as plant growth substances, impinge upon the genetic regulatory mechanism of a plant by affecting one of the intermediate
reactions involved in transcription of a particular DNA at a
particular time in the life cycle of a cell?" One would antici-
pate that the probability of locating the site of action of a
specific external stimulus or of a plant growth regulator would
be highest if the suspected macromolecules and intracellular
organelles of the model system were derived from the tissue in
which the regulator exerts its control, i.e., exclusively from
higher plant parts.

Success in the search for RNA polymerase in eukaryotes was
first attained by Weiss [1]. He examined rat liver nuclei for
ATP-incorporating activity that was stimulated by the addition
of the triphosphates of the other three nucleotides found in RNA.
Bonner early demonstrated RNA polymerase in plants when he and

his colleagues [2] purified a polymerase from the chromatin
fraction of pea embryos. In fact, my early attempts to enhance
the polymerase activity in corn chromatin were conducted with
Huang and Bonner while I was visiting at the California Institute of Technology. Early attempts to purify the polymerase
from corn were subsidiary to Novelli's and my main experimental
thrust, which was to provide a source of homologous messenger
RNA to stimulate a cell-free amino acid-incorporating system
from corn seedlings [3]. Only after we became involved in the
isolation of the polymerase did understanding of the transcription process become an experimental objective unto itself.

The selection of maize seedlings as an enzyme source de-
serves discussion. The germinated corn grain is a eukaryotic
tissue rich in genetic markers [4] and is a potentially available source of gene-control systems [5]. Because of the extensive hybrid corn breeding program in this country, maize
genetics is among the most extensively studied of the higher
plants. The germinated grain presents the investigator with a
mosaic of cell types at various stages of differentiation and
parental genetic information (diploid maternal pericarp, diploid
embryo, triploid endosperm). Furthermore, a homogeneous population of one cell type somewhat analogous to tissue cultures
can be obtained by gross dissection. Ready availability of
grain and ease of germination of large batches of material in
the laboratory meet the enzymological prerequisites for an

enzyme source. The responsiveness of protein- and RNA-synthe-
sizing systems isolated from the rapidly growing tissue to en-
vironmental stimuli [6] makes corn seedlings an experimentally
rich tissue for isolation of transcription machinery and medi-
ating control components.

It had been anticipated and found that DNA-dependent RNA
polymerases were associated with the chromatin fraction of
eukaryotes, closely associated with their required templates
[1, 2]. Furthermore, the preparation of microbial RNA polymerases from Escherichia coli [7], Micrococcus lysodeikticus
(luteus) [8], and Azotobacter vinlandii [9] required coprecipi-
tation of the polymerase with cellular nucleic acids, then
selective elution of the active protein from the nucleoprotein
aggregates. In early reports of the work with chromatin fractions derived from plants [2], the soluble proteins apparently
were not examined for polymerase activity, or, if examined,
were found to be inactive. In marked contrast, the enzyme
activity tentatively identified as RNA polymerase was present
in the soluble fraction dissociated from the insoluble chromatin
of corn homogenate [10]. Therefore, we embarked upon its purification. The occurrence of the nucleotide-incorporating activity
in a soluble fraction relatively free of DNA facilitated early
recognition of the DNA requirement, which had been undemonstrable
with enzymes from eukaryotic tissues.

Classical enzymological techniques were, therefore, directly applicable to the isolation and purification of the active
protein from corn seedlings. When subjected to techniques of
isolation and purification of cellular components from tissues
other than plants, a corn seedling behaves more nearly like an
animal than like Escherichia coli, not a surprising result since
it is eukaryotic. Nevertheless, the methodology developed for
the isolation of microbial polymerases constitutes the basis for
many of the procedures used in our laboratory for the last 15
years.

III.   ISOLATION AND PURIFICATION

A.   Germination and Harvesting

Early attempts to isolate RNA polymerase from corn seedlings
were strongly influenced by experience gained in the isolation
of ribosomes and polysomes active in amino acid incorporation [3].
Germination of corn grain (Zea mays L., WF9 X Bear 38, waxy)
in 10- to 100-kg batches in perforated plastic pails under a
constant shower of tap water at 23°C was modified from the procedure described by Huang and Bonner [11]. Gentle shaking,
three times daily, prevents packing of the swelling grain to
assure adequate aeration and temperature regulation for uniform
germination and to facilitate physical removal of microbial contaminants with running water. At lower temperatures more root
than shoot tissue emerges, and above 25°C, nuclease levels in

homogenates prepared from the shoots are inordinately high.
Shoots from seedlings germinated in vermiculite are laden with
mold spores and bacterial contaminants and are of no use. Shoots
of grain germinated on trays lined with filter paper (Whatman
number 1 may be assumed to be sterile)are useful, but the yield
is too limited for enzyme preparation. After four or five days
germination, or when the shoots are 1 to 2 cm long, they are
separated from the grain and most of the roots by rubbing the
seedlings on a stack of vibrating, stainless steel screens
flooded with cold tap water. We use a stainless steel gravel
separator (Sweco, 18 in. diam) although manually rubbing the
seedlings on a stainless steel 4-mesh (0.0475 wire diam, 0.2023
opening) beneath running water works almost as well (but is
fatiguing). To minimize decay of polymerase activity in the
detached tissue, the shoots are collected in a hand strainer,
immediately frozen in liquid nitrogen, and--after harvesting
is completed--stored in 90-g packets at -76°C. Plucking individual plumules (shoots) into liquid nitrogen is feasible
for use in detection of activity, but is too time-consuming
for an enzyme preparation. With the separator, 13 kg of shoots
are harvested and packaged in less than 2 hr. The frozen shoots
may be ground under liquid nitrogen to a powder before storage
at -20°C and the polymerase remains active for a month or more.
Thawed shoots yield inactive preparations.

B. Homogenization

The homogenizing medium, 100 mM Tris·HCl, pH 7.6 at 23°C
(pH 8.0 at 4°C), 1 mM $MgCl_2$, 50 mM 2-mercaptoethanol, and 250 mM
sucrose), was adapted from that used in the isolation of active
polysomes from corn seedlings [3] and that of Huang and Bonner
[11]. All the components are essential for the isolation of
high levels of nucleotide-incorporating activity in the homogenate. The Tris concentration must be 50 mM or higher and the pH
above 7.8 to maintain the homogenate above pH 7.5; at acid pH
values polymerase activity falls precipitously. In the absence
of a sulfhydryl reagent the extract turns dark brown within
minutes (presumably from the accumulation of oxidation products
generated by peroxidases) and is inactive in nucleotide incorporation, while the proteins behave unpredictably upon salt
precipitation. Dithiothreitol is less effective than 2-mercap-
thanol, probably because the oxidized dithiothreitol generated
by the extract is insoluble and is also inhibitory to nucleotide incorporation. The role of metal ions is unknown, but they
may be involved in keeping the RNA polymerase-nucleic acid complex intact during the initial step in polymerase isolation.
Omission of magnesium results in insoluble pellets after ammonium
sulfate precipitation. Maintenance of intact ribosomes by inclusion of divalent cations also facilitates their subsequent
removal. Although used on occasion [13], 0.1 mM EDTA is not
required in the homogenizing medium; however, addition to re-
suspension media [14] used in subsequent steps of EDTA does limit

apparent nuclease activity. The addition of sucrose to the medium stabilizes the extract for subsequent differential centrifugation by virtue of the increased viscosity of the extract, and seems to stabilize the polymerase complex when it is separated from the DNA. Since essentially all organelles are disrupted by the homogenization procedure, maintenance of the proper tonicity of the medium with sucrose for membrane integrity probably is irrelevant. Medium is prepared just prior to use by combining aliquots of concentrated reagents (usually 1 M solutions frozen in appropriate small volumes), solid sucrose, and freshly distilled water. Use of fresh medium minimizes contamination of the homogenate by molds and bacteria that may germinate and grow in refrigerated reagents.

Three methods of homogenization have been used with success: a French pressure cell, a ground-glass homogenizer, and a Waring Blendor. Initially [3] the polymerase was isolated as a soluble enzyme from homogenates prepared by passage of a slurry of powdered shoots (ground under liquid nitrogen with an electric mortar and pestle) and homogenizing medium (1 g powder/1.5 ml medium) through a French pressure cell (11,000 psi at 0°C). We erroneously ascribed the release of the polymerase from the bulk of the DNA in the corn homogenate, as compared to the chromatin-bound enzyme in pea and rat liver nuclei, to this method of homogenization. In experiments using five- or seven-day old corn shoots, we found that the RNA polymerase was in the

soluble compoment of homogenates prepared with a loose-fitting
glass homogenizer [12]. The shoots were frozen in liquid nitrogen on harvesting, the glass homogenizer was maintained at 0°C,
and the resultant homogenate was a viscous slurry. We have
found that highly active polymerase preparations can be most
easily obtained with a Waring Blendor [15]. To a Pyrex homogenizer bowl (at 23°C) are added the frozen stems of a 90-g packet
and 135 ml of fresh medium (at 23°C). The bowl and the medium
must not be prechilled, because addition of the stems frozen at
-76 °C will then freeze the mixture solid. The mixture is homogenized at low speed for 1 min or until a cornmeal-colored,
stiff slurry, which circulated without stirring, is obtained.
The slurry is immediately filtered through four layers of gauze
through a chilled funnel into a cold flask. The retained pulp
is squeezed by twisting the free ends of the gauze so as to
constrict the bolus; grasping the bolus must be avoided (to
avoid heating the extract) and plastic gloves must be worn (to
avoid introducing "finger" nuclease). The gauze filtrate is
immediately passed through one layer of Miracloth (glass paper)
to remove the lipid-rich membraneous material, and the filtrate
is immediately distributed into chilled centrifuge tubes. The
entire procedure is accomplished in less than 5 min and can be
performed on a laboratory bench with ice buckets used to chill
glassware and filtrates. The level of polymerase detected in
the homogenate is inversely related to the time elapsed between

homogenization of the shoots and centrifugation of the filtered
homogenate.

C. Centrifugation

The filtered homogenate is centrifuged at 50,000 rpm for
60 min in a Ti50 rotor (200,000 x g) at 0°C. The filtrate obtained from 90 g of shoots (150 ml) is accommodated by this
rotor (144 ml); in fact, the weight of shoots per packet is
determined by the rotor capacity. Larger batches can be accommodated with rotors of greater capacity; however, since running
time must be increased for comparable precipitation times with
larger rotors, runs longer than 2 hr should be avoided. Centrifugal schedules adequate to remove ribosomes will yield soluble
components with RNA polymerase activity [14]. DNA-dependent
nucleotide-incorporating activity is detected in soluble components prepared at lower centrifugal forces, i.e., in a Sorvall
centrifuge at 20,000 x g [3]. However, the presence of ribosomal RNA and proteins interferes with subsequent salt precipitation so as to make removal of nucleases difficult. Intermediate centrifugation before removal of ribosomes is not recommended.
The clear, straw-colored soluble component is decanted from the
greenish-brown pelleted chromatin, ribosomes, and cellular
organellar debris through Miracloth into a chilled beaker. If
the supernatant solution is cloudy the homogenate may have been
ground for too long, the preparation may have been warmed, or
work may have proceeded too slowly. It may be necessary to

increase the buffer or sulfhydryl concentration in the medium.
In any case, it is necessary to start again. If a lipid pellicle
is present on top of the centrifuged homogenate and it is not
removed from the soluble component by filtration through two
layers of Miracloth, then the germination time of the seedlings
must be lengthened to further exhaust the stored lipid in the
grain or to germinate a less fatty grain. Visible amounts of
lipid will interfere with subsequent fractionation steps.

D. Salt Precipitation

Several objectives are accomplished by selective precipitation of proteins in the high-speed supernatant with ammonium
sulfate. The relatively dilute soluble proteins (2-4 mg/ml)
are concentrated (to 20-30 mg/ml) upon resuspending the precipitated proteins in a near-minimal volume. A major portion of
the nuclease activity is left behind among the unprecipitated
proteins [14]. The sucrose and magnesium added with the homogenizing medium are drastically reduced. Selective precipitation
between 25% and 40% saturation with ammonium sulfate was used
in the initial isolation when homogenates were prepared with a
French pressure cell [14]. Solutions of soluble proteins obtained as high-speed supernatants from homogenates prepared
with a Waring Blendor are routinely brought to 50% saturation
by the addition of an equal volume (130 ml) of saturated
ammonium sulfate. The saturated ammonium sulfate is stored at
0°C and is adjusted to pH 8 (dilute 1:20) with ammonium hydroxide just before use (be certain undissolved salt is present).

The salt solution is added slowly (4 to 5 min) to the supernatant component with constant but slow stirring (magnetic stirrer
fitted with an ice bath) at 0°C. Stirring is continued for 30
min, then the mixture is decanted into chilled plastic tubes and
is centrifuged at 10,000 x g for 10 min. The clear-yellow supernatants are decanted, the pellets are drained, and the walls of
the inverted tubes are wiped to remove lipid and residual supernatant solution. Cold resuspending medium (50 mM Tris·HCl, pH
8.0 at 4°C, 10 mM 2-mercaptoethanol; 0.5 ml) is added to each
pellet, and the pellet is readily worked into solution with a
rubber policeman; finally, all pellets are pooled, along with a
serial wash of the tubes, and brought to 12.5 ml with cold resuspending medium in a chilled, calibrated tube. The resuspended
protein solution should look clear in a 1-ml pipet. If the
solution is cloudy, add 1-2 ml more of resuspending medium. If
the solution fails to clear, the proteins may have been heat-
denatured or all the lipid on the walls may not have been wiped
away; discontinue the preparation and start again. The lower
portion of a tube should not be held with one's fingers (36°C),
and chilled or plastic pipets should be used to transfer the
resuspended proteins and the medium to minimize heat denaturation.

E. Desalting

Removal of salt from the resuspended ammonium sulfate pellets
is a critical step in the purification scheme. The two objectives are to remove sufficient salt to permit assay of polymerase

and to adjust the salt concentration to permit selective adsorption to DEAE-cellulose in the subsequent step. Preparations derived from French pressure-cell homogenates were dialyzed for 4 hr against resuspending medium at 0°C [14]. We now routinely pass the resuspended proteins (12.5 ml) over a Sephadex G50 column (2.5 cm i.d. x 11 cm high) equilibrated with 50 mM Tris·HCl pH 8.0, 10 mM 2-mercaptoethanol, 0.20 M ammonium sulfate (equilibrating medium) at 0°C. The gel filtration procedure takes less than 45 min, rids the preparation of low molecular weight (10,000 or less) contaminants, and establishes the salt concentration at 0.20 molar. More than 90% of the added polymerase activity is recovered in the 16-ml excluded volume. The excluded volume may be anticipated by collection of a few additional milliliters, but collection should be discontinued before elution of included material. The sample volume may be expanded to 25 ml to facilitate absorption in the subsequent step. We prefer Kontes Glass Chromaflex or Pharmacia plastic columns to obtain minimum dilution of the excluded volume. Calibration of a thoroughly equilibrated column (pass 200 ml of equilibrating medium) with a mixture of Dextran 2000 and $[^{14}C]$leucine equivalent to the volume of the protein to be desalted (12.5 ml) is convenient and avoids confusion as to when the yellow excluded protein is eluted from the column and when the retained yellow pigments and salt are eluted.

F. DEAE-Cellulose Chromatography

The major resolution of RNA polymerase from the other soluble proteins is accomplished by DEAE-cellulose column chromatography. Three definite objectives are achieved by this one procedure: (1) selective absorption of the polymerase to the cellulose, with the exclusion of contaminating nucleases, NTP:polynucleotidylexotransferase [16] and polynucleotide phosphorylase [14]; (2) selective elution of the polymerase in a small volume by a steep, linear salt gradient; and (3) complete dissociation and removal of the polymerase from residual nucleic acids. DEAE-cellulose--washed, sized, and cycled with acid and alkali in 250-g batches as described by Peterson and Sober [17]--is stored at 4°C. Although manufacturers claim otherwise, we find it advantageous to buy a large supply of one lot of resin to insure reproducibility over extended purification procedures. We rewash and use the DEAE-cellulose three times. A day before use, a column 2.5 cm i.d. x 11 cm high is poured, packed by gravity flow, and equilibrated with about 150 ml of equilibrating medium at 4°C. By use of a four-veined Buchler peristaltic pump the flow rate over the column is maintained constant, the gradient is mixed and delivered linearly, and the volume delivered to the column is monitored continuously. There are five components: (1) the column, (2) a 0.2 M salt reservoir, fitted with a magnetic stirrer and open to the atmosphere, (3) a 1 M salt reservoir open to the atmosphere, (4) a graduated 100-ml

cylinder, and (5) a continuous-flow cuvette with a 2-mm light
path, all of which are interconnected with four lengths of Tygon
tubing (1/32-in. i.d. x 1/8-in. wall). The connections are:
(1) from the 0.2 M salt reservoir through the pump to the top
of the open column, (2) from the 0.2 M salt reservoir through
the pump to the graduated cylinder, (3) from the 1 M salt reservoir through the pump to the 0.2 M salt reservoir, and (4) from
the bottom of the column through the pump to the continuous flow
cuvette. The eluate is led through a continuous-flow ultraviolet-
monitoring system (we use a Gilford spectrophotometer) and the
A_{280} is recorded. Once established, the reproducibility of the
elution profile makes use of the ultraviolet monitoring system
unnecessary, and enzyme-bearing fractions can be identified by
the elution volume collected in the cylinder. Before application of the sample, the column is equilibrated (by use of the
pump) with an additional 50 - 100 ml of equilibrating medium
(after it is allowed to remain overnight in the cold). The
25-ml sample of excluded proteins is applied to the DEAE-cellulose column at a rate of 1 ml/min (pump flow rate) and washed on
with an additional 50 ml of equilibrating medium; the effluent
containing the unabsorbed proteins (80-85% of the added proteins)
is collected in a graduated cylinder. The flow rate is reduced
to 0.25 ml/min and the gradient is started (60 ml of 0.20 M
ammonium sulfate and 60 ml of 1 M ammonium sulfate, both in 50
mM Tris·HCl, pH 8.0, 10 mM 2-mercaptoethanol; reagents are prepared

from distilled water at 23° to minimize formation of bubbles
from dissolved air). Fractions (2.5 ml) are collected directly
into screw-cap vials and frozen in liquid nitrogen. Although
the enzyme is stable to repeated rapid freezing and thawing,
it is labile at 4°C. Polymerase appears after 10 ml of the
gradient has been collected and is eluted in the next 7.5 - 12.5 ml.
The ultraviolet profile of the proteins absorbed and then eluted
from the column is a reliable indication of the polymerase
elution position; that segment of the elution profile is presented
in Fig. 1. Although a linear gradient is delivered to the
column, the salt concentration at various stages during elution
must be determined empirically with a refractometer; the salt
gradient plotted in Fig. 1 was derived from refractometer read-
ings corrected for protein. We were able to purify the polymerase 100-fold by gradient elution from DEAE-cellulose [14] with
a 50 mM to 1 M Tris gradient, but we failed to recover activity
when elution was attempted with ammonium sulfate. Equilibration
of the column with 0.20 M ammonium sulfate coupled with the steep
gradient enabled us to recover 70% of the α-amanitin-sensitive,
AMP-incorporating activity detected in the 200,000 x g supernatant
as a 200-fold purified RNA polymerase II. Since the specific
activity detectable in the filtered homogenate is less than 0.05
nmoles of AMP/mg protein in 20 min, this procedure results in a
net purification of greater than 1000-fold. The polymerase is
not completely absorbed to DEAE-cellulose equilibrated with

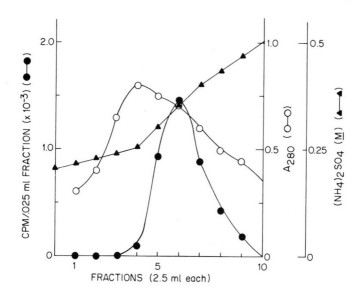

Fig. 1   Elution of RNA polymerase from DEAE-cellulose.

higher than 0.20 M ammonium sulfate.   Batch elution with 0.4 M
ammonium sulfate is more rapid and yields a more concentrated,
but less homogeneous, polymerase preparation.   From homogenization of the shoots through elution of the polymerase from the
DEAE-cellulose column, the procedure takes 11 to 12 hr, provided
columns and reagents are prepared the previous day.   Fractions
are obtained at intermediate stages, frozen in liquid nitrogen,
and later assayed for enzymatic activity and protein; the
efficacy of each of the steps in the fractionation is monitored
by these analyses.   Values obtained should fall within 10 - 15%
of those presented in Table 1.

TABLE 1

Purification of RNA Polymerase

Fraction	Volume (ml)	Total protein (ml)	Total activity (units)[a]	Specific Activity[b] (units/mg)
200,000 x g Supt.	130	500	250	0.50
50% $(NH_4)_2SO_4$ Ppt	12.5	270	162	0.60
Sephadex G-50 excluded	25	240	110	0.46
DEAE-cellulose fraction #5	2.5	0.80	34	42
#6	2.5	0.50	52	104
#7	2.5	0.33	32	98
#8	2.5	0.22	8.8	38
#9	2.5	0.17	6.3	37

[a]Unit is equivalent to 1 nmole of AMP incorporated at 30°C in a standard assay mixture (see Table 2).

[b]Specific activity is units/mg of protein at 20 min.

IV. ASSAY

The presence of multiple RNA polymerases in cells of higher plants [18-21] was initially demonstrated in mammalian cells by Roeder and Rutter [22]. Detection of multiple poly- merases, as well as other nucleotide-incorporating enzymes, in plants [16, 23, 24] can be ascribed to manipulation of the salt, metal, and template requirements of the RNA polymerases

and the judicious use of antibiotic inhibitors in assay of
their activity. Organellar isolation before enzyme isolation
has been successful [21], but I believe that our current technology can be most efficiently taken advantage of by resolving
macromolecules with enzymatic activity from completely disrupted
cells and identifying them as polymerases and proceeding with
their purification. By organellar isolation these activities
can then be localized within the compartments of a cell, as
done by Blatti et al. [25].

A. General Considerations

Since the product of all RNA polymerases, by definition,
is RNA, it is axiomatic that the assay must measure the accumulation of RNA. Since in the process of transcription the sequence of nucleotides in 3', 5'-phosphodiester linkage must be
determined by the nucleotide sequence of DNA, not only are all
four nucleotides required as precursors, but DNA must be required
as a template for base alignment by Watson-Crick pairing. Thermo-
dynamic considerations of polymer synthesis demand that nucleoside di- or triphosphates must be the immediate precursors. The
specificity of the RNA polymerases is for the triphosphates, in
contrast with the nontemplate-requiring polymerization of nucleoside diphosphates catalyzed by polynucleotide phosphorylase [26].
Like most kinase reactions, a divalent metal ion is required
for nucleotidyl transferase activity. Therefore, presumptive
evidence for RNA polymerase is the requirement for DNA (native

or single-stranded), for a metal ion (magnesium or manganese),
and for the accumulation of acid-insoluble product from a mixture of ATP, CTP, GTP, and UTP. By use of one of the nucleoside
triphosphates as a labeled precusor, a requirement for the other
three nucleoside triphosphates for the accumulation of acid-
insoluble material (oligomers at least five or six nucleotides
long) constitutes the basis for polymerase assays. Increased
specific activity of commercially available isotopes and increased counting efficiency of liquid scintillation spectro-
meters permit the detection of less than picomole levels of
nucleotide incorporated. However, hypersensitive assays also
increase the possibility of detection of acid-insoluble incorporation into products other than RNA [23] by entities other
than RNA polymerase [16]. Use of filter paper disks or Millipore
filters facilitates assay of enzymatic activities and obviates
the laborious washing of individual acid-insoluble precipitates
[27]. We prefer the use of filter paper disks. However, if
tritiated substrates are used, Millipore filters give much
higher counting efficiencies, 25% on Millipore filters versus
6% on paper [28]. Although tritiated substrates are less expensive than ^{14}C- or ^{32}P-labeled nucleotides, the cost of
Millipore filters and additional trichloroacetic acid as compared with filter paper disks negates the savings accrued in
the purchase of tritiated substrates.

B. Specific Assay Conditions

The assay of RNA polymerase II from maize seedlings consists of incubation of an aliquot of an enzyme preparation with
radioactive ATP, cold CTP, GTP, and UTP, calf-thymus DNA, and
magnesium in Tris buffer with 2-mercaptoethanol at 30°C, removal
of aliquots after 20 min of incubation, and determination of
the acid-insoluble radioactivity accumulated in a 0.1-ml incubation mixture. A protocol is presented that utilizes a minimum of reagents at or near their optima and accommodates samples
that vary widely in protein concentration (and volume if necessary), as well as that permits addition of activators (salt or
factors) and specific inhibitors (α-amanitin). The results of
one assay are presented in Table 2. Although obvious to "workers
in the field," there are several tricks, short-cuts, and cautions
that should be indicated to the reader to assure success in
assaying for RNA polymerase II.

Careful preparation, storage in small aliquots, and avoid-
ance of cross-contamination of the concentrated and expensive
reagents are essential. Tris, purchased from Sigma, is prepared by mixing 1 M Tris base and 1 M Tris·HCl to pH 8 at 23°C,
it is stored in 1-ml aliquots at -15°C. If the buffer is not
colorless (i.e., yellow) discard it. Concentrated 2-mercapto-
ethanol (14.6 M) is diluted to 0.4 M just before use; the diluted
reagent if stored at -15°C is auto-oxidized and becomes inhibitory for polymerase. The nucleoside triphosphates (previously

TABLE 2a

Assay of Fractions in Preparations of Maize RNA Polymerase II

Reaction Mixture

Reagent	ml/assay tube	ml to pipet	final conc
1 M Tris·HCl, pH 8.0 at 25°C	0.01	0.20	100 mM
0.4 M 2-mercaptoethanol	0.001	0.02	4 mM
0.2 M MgCl$_2$	0.01	0.20	10 mM
25 mM each of CTP, GTP, UTP (sodium)	0.005	0.10	2.5 mM(each)
0.071 M ATP - Na	0.0015	0.03	1.1 mM
90 µCi/ml [8-^{14}C]ATP (24 Ci/mole)[a]	0.0025	0.05	2.25 µCi/ml
37 A$_{260}$/ml Calf-thymus DNA	0.015	0.30	5.55 A$_{260}$/ml
10 mg/ml Bovine serum albumin	0.005	0.10	50 µg/ml

[a]ATP final specific activity = 1.69 Ci/mole; equivalent to 2.78 cpm/pmole of AMP.

stored as powders at -15°C in a CaCO$_3$—charged mason jar) are prepared individually by solution in cold distilled water, and the pH of the resultant solutions is adjusted to 7.0 with NaOH (never exceed pH 7 or terminal phosphates will be hydrolyzed). The concentrations are determined spectophotometrically and the three unlabeled nucleotides are mixed and stored in 0.5-ml volumes at -15°C. The fourth nucleoside triphosphates (ATP in the example given here) are not included in the mixture, so that the final activity, as well as the final concentration, of the labeled precursor can be adjusted in the incubation mixture. Radioactive nucleotides should be counted and chromatographed before use to insure activity and purity upon receipt,

Reaction Mixture

Reagent, µl	1	2	3	4	5	6....	13	14	15	16....	18	19
Reaction mixture	50	50	50	50	50	50	50	50	50	50	50	50
H_2O	20	15	25	20	25	20	25	25	25	25	20	50
M $(NH_4)_2SO_4$	5	5	-	-	-	-	-	-	-	-	-	-
α-amanitin (1 mg/ml)	-	5	-	5	-	5	-	-	-	-	-	-
200,000 x g supt.	25	25	-	-	-	-	-	-	-	-	-	-
50% $(NH_4)_2SO_4$ ppt.	-	-	25	25	-	-	-	-	-	-	-	-
Sephadex G50 excluded	-	-	-	-	25	25	-	-	-	-	-	-
DEAE-cellulose fract #5	-	-	-	-	-	-	25	-	-	-	-	-
fract #6	-	-	-	-	-	-	-	25	-	-	-	-
fract #7	-	-	-	-	-	-	-	-	25	-	-	-
fract #8	-	-	-	-	-	-	-	-	-	25	-	-
Known polymerase	-	-	-	-	-	-	-	-	-	-	25	-
cpm/disk - background (29.5)	134	18	911	234	309	76	98	938	1458	896	285	0
Enzyme protein added (µg)	96	96	540	540	240	240	9.0	8.0	5.0	3.25	11.2	0
Specific activity (nmoles AMP/mg of protein)	0.5	0.07	0.6	0.15	0.46	0.11	3.9	42	104	98	9.1	-

and after storage as well. Ethanol present in the isotopic precursor must be removed or reduced to 5% (azeotropic mixtures are
permissible with labeled nucleotides of 20 Ci/mole or higher,
since greater dilution before use is possible). Evaporation at
0°C under a fine stream of filtered air of an aliquot delivered
to a plastic tube (that tube used to prepare the reaction mixture)
is rapid and avoids large losses of isotope on glass surfaces.
Commercially available, high molecular weight, calf-thymus DNA is
dissolved overnight in 0.1 x SSC (15 mM NaCl, 1.5 mM sodium citrate, pH 7.3) and frozen in 0.2-ml aliquots (approximate volume
per assay). We have noted a transient increase in template
activity of stored DNA, consistent with the preference of the
maize polymerase for denatured DNA [29]. Reagents are thawed
just before use, held on ice during preparation of the reaction
mixture, and refrozen immediately after use.

A reaction mixture is prepared containing all reagents
common to each incubation mixture, so as to minimize pipetting
errors within each assay. Since the assay involves the mixture
of many [12] concentrated reagents in a small volume (0.1 ml),
the order of addition (buffer first, metal ion before DNA, and
isotope last--except when ethanol must be removed) as presented
in the protocol avoids formation of insoluble or inactive complexes. If $MnCl_2$ is substituted for $MgCl_2$, the $MnCl_2$ should be
less than 50 mM and should be added just before the isotope, to
a final concentration of 5 mM. Should the mixture turn a dirty

brown and fail to clear within a minute, reduce the final concentration of the $MnCl_2$ further. Thorough mixing by rolling the
tube so as to slosh the solution over any part of the inner wall
of the tube that has come in contact with the reagents (do not
mix by shaking) enhances activity and reproducibility of results.
A reaction mixture adequate for one more incubation tube than is
required for assay is prepared. The additional volume permits
counting of an aliquot to determine the radioactivity actually
added to the assay, and allows one to correct for any losses
incurred in pipetting and surface adsorption. Disposable plastic
tubes (12 mm x 75 mm) are particularly useful because of their
surface properties and low heat transfer. Disposable micropipets
are strongly recommended for dispensing reagents into the reaction
mixture, dispensing the mixture into assay tubes, and removing
aliquots for assay.

The protocol is usually limited to 19 tubes to permit starting and stopping 20-min incubations at 1-min intervals. An experienced "manipulator" can easily manage 39 tubes at 30-sec
intervals. If aliquots of the incubation mixture are to be removed at intermediate times, the number of tubes incubated must
be adjusted accordingly. To chilled plastic tubes, 0.05 ml of
the reaction mixture is added, as well as water (necessary to
adjust the final volume of the completed incubation mixture to
0.1 ml), and additional activators and/or inhibitors; the tubes
are held on ice. The reaction is initiated by addition of enzyme

and incubation at 39°C in a water bath. No change in rate of
reaction is detected by warming the incubation tubes before enzyme
addition. Since the soluble RNA polymerase from maize requires
50 - 100 mM salt for activity [14], ammonium sulfate is added to
the high-speed supernatant assay tubes. All other fractions except the precipitate suspended in 50% ammonium sulfate (salt
concentration is unknown and assay of activity is variable) are
in 0.2 M ammonium sulfate. Since nucleotide-incorporating activities other than RNA polymerase II are present in intermediate
fractions before DEAE-cellulose chromatography [13], a companion
incubation mixture containing α-amanitin is included to facilitate
calculation of RNA polymerase II activity in the intermediate
fractions (assuming all α-amanitin-sensitive nucleotide incorporation is RNA polymerase II [22], which may not be true. We have
found that poly(A) synthesis catalyzed by this enzyme preparation
is sensitive to α-amanitin [15]). Included with the assay of
preparations of unknown activity are two incubation mixtures,
one with an enzyme of known specific activity and the other without enzyme. The first provides data for calibration of the
assay. The second is a monitor of the wash-up procedure and
should be at background. After 20 min of incubation, the tube
contents (0.1 ml) are pipetted on to a pin-mounted filter-paper
disk, held for 15 sec under a hair-dryer to hasten absorption
into the paper, and dropped into a plastic basket immersed in
400 ml of cold 10% trichloroacetic acid--2 mM sodium pyrophosphate

(the pyrophosphate minimizes nonspecific adsorption of the unincorporated radioactive nucleotide). Ten minutes after collection of the last sample the disks are transferred, in the
basket, to 10% trichloroacetic acid-2 mM pyrophosphate and extracted, with vigorous stirring beneath the basket, for 5 min
at 23°C. Three additional extractions are performed. The disks
are then extracted at 37°C for 5 min with ethanol-ether (1:1)
to remove lipid, water, and trichloroacetic acid, washed with
ether to remove the ethanol, and finally dried under a heat
lamp to rid the disks of the ether. Dried disks are counted
in 5 ml of dimethyl POPOP-toluene based scintillator (after
counting, disks are removed, the scintillator solution is replaced, the vials are monitored, and those that are not contaminated may be reused).

An aliquot of each enzyme preparation is precipitated with
10% trichloroacetic acid, the precipitate is dissolved in 0.1 M
NaOH, and the amount of protein is determined colorimetrically
by the method of Lowry et al. [30] with bovine serum albumin
as a standard. Protein in fractions eluted from the DEAE-cellulose column containing polymerase may be estimated from the A_{280},
but extinction coefficients do vary across the column and the
Lowry determination is recommended wherever possible. Calculation of the final enzyme specific activity at 20 min of incubation is from the raw counts accumulated, the total counts added,
the amount of ATP added (specific activity of the precusor and

the counting efficiency of the disks at 75%), and the amount of

protein added to the incubation mixture.

Failure to detect enzymatic activity (with polymerase of

known activity) is most often due to an error in arithmetic!

Check all calculations, reagent concentrations, dilutions, and

volumes pipetted with a sympathetic friend. If no error in

calculations was responsible, then attempt to perform a positive

assay by: (1) increasing the specific activity of the radio-

active precusor to at least 10 Ci/mole, (2) varying the salt

concentration from 25 to 150 mM, (3) reducing the amount of

protein added of the intermediate fractions to 50 - 80 µg per

incubation tube, or increasing the amount of protein of the

DEAE-cellulose fractions to 5 - 10 µg per incubation mixture,

(4) using $MnCl_2$ in place of $MgCl_2$ (with appropriate changes in

concentration as noted above); (5) using heat-denatured DNA or

poly(dT) as the template, and finally (6) if all else fails,

making   arrangements to visit a laboratory that publishes active

polymerase assays.

The assay with appropriate modifications will detect sev-

eral nucleotide-incorporating activities. We have recently

adapted the assay for rapid scanning of all fractions eluted

from the DEAE-cellulose column. Aliquots (25 µl) of the re-

action mixture are delivered to depressions in a warm (30°C)

glass spot-plate. An aliquot (25 µl) of a given fraction is

added. After 20 min of incubation, the contents of the depression are absorbed to a folded filter paper disk. The disk is dropped into 10% trichloroacetic acid--2 mM sodium pyrophosphate and washed as described above. Although not as precise or reproducible as the tube assay, the spot-plate assay permits rapid screening of many samples for several nucleotide-incorporating activities. No additional elaborate or expensive instrumentation is required.

V. CONCLUDING REMARKS

I have described in some detail the isolation, purification, and assay of RNA polymerase II from corn seedlings. The enzyme is not yet homogeneous and we have not identified or purified each of the polypeptide subunits constituting the active complex. Other nucleotide-incorporating activities in the soluble component of the seedling homogenate--as well as those associated with the chromatin, other particulates, and organelles--remain to be identified and purified. Finally, relationships among the various RNA polymerases, as well as their structural and functional relationships (if any) with DNA polymerases and reverse transcriptase(s), demand our attention.

REFERENCES

[1] S. B. Weiss, Proc. Nat. Acad. Sci. USA, 46, 1020 (1960).

[2] J. Bonner, R. C. Huang, and N. Maheshwani, Proc. Nat. Acad. Sci. USA, 47, 1548 (1961).

[3] R. J. Mans and G. D. Novelli, Biochim. Biophys. Acta, 80, 127 (1964).

[4]   J. Weijer, Bibliographia Genetica, XIV, 189 (1952).

[5]   B. McClintock, Amer. Naturalist, 95, 265 (1961).

[6]   R. J. Mans, in Biochemistry of Chlorophasts (T. W. Goodwin, ed.), Vol. II, Academic Press, London, 1966, p. 351.

[7]   M. Chamberlin and P. Berg, Proc. Nat. Acad. Sci. USA, 48, 81 (1962).

[8]   T. Nakamoto, C. F. Fox, and S. B. Weiss, J. Biol. Chem., 239, 167 (1964).

[9]   S. Ochoa, D. P. Burma, H. Krooger, and J. D. Weill, Proc. Nat. Acad. Sci. USA, 47, 670 (1961).

[10] R. J. Mans and G. D. Novelli, Biochem, Biophys, Res. Commun., 91, 186 (1964).

[11] R. C. Huang and J. Bonner, Proc. Nat. Acad. Sci. USA, 48, 1216 (1962).

[12] E. R. Stout, R. Parenti, and R. J. Mans, Biochem. Biophys. Res. Commun., 29, 322 (1967).

[13] T. J. Walter and R. J. Mans, Biochim, Biophys, Acta, 217, 72 (1970).

[14] E. R. Stout and R. J. Mans, Biochim. Biophys. Acta, 134, 327 (1967).

[15] R. H. Benson and R. J. Mans, Fed. Proc., 31, 427 Abs (1972).

[16] R. J. Mans and T. J. Walter, Biochim. Biophys. Acta, 247, 113 (1971).

[17] E. A. Peterson and H. A. Sober in Methods in Enzymology (S. P. Colowick and N. O. Kaplan, eds.), Vol. III, Academic Press, New York, 1957, p. 785.

[18] P. A. Horgen and D. H. Griffin, Proc. Nat. Acad. Sci. USA, 68, 338 (1971).

[19] H. Mondal, R. K. Mandal, and B. B. Biwas, Biochem. Biophys. Res. Commun., 40, 1194 (1970).

[20] G. C. Strain, K. P. Mullinix, and L. Bogorad, Proc. Nat. Acad. Sci. USA, 68, 2412 (1971).

[21] W. Bottomley, H. J. Smith, and L. Bogorad, Proc. Nat. Acad. Sci. USA, 68, 2412 (1971).

[22] R. G. Roeder and W. J. Rutter, Nature, 224, 234 (1969).

[23] R. J. Mans, Biochem. Biophys. Res. Commun., 45, 980 (1971).

[24] C. T. Duda and J. H. Cherry, J. Biol. Chem., 246, 2487 (1971).

[25] S. P. Blatti, C. J. Angles, T. J. Lindell, P. W. Morris, R. F. Weaver, F. Weinberg, and W. J. Rutter, Cold Spring Harbor Symp. Quant. Biol, 35, 649 (1970).

[26] Y. Kimhi and U. Z. Littaner, in Methods in Enzymology, (L. Grossman and K. Moldare, eds.), Vol. XII, Academic Press, New York, 1968, p. 513.

[27] F. J. Bollum, Progress in Nucleic Acid Research (Cantoni and Davies, eds.), Harper and Row, 1966, p. 296.

[28] R. J. Mans and G. D. Novelli, Arch. Biochem. Biophys., 94, 48 (1961).

[29] E. R. Stout and R. J. Mans, Plant Physiol., 43, 405 (1968).

[30] O. H. Lowry, N. J. Rosebrough, A. L. Farr, and R. J. Randall, J. Biol. Chem., 193, 265 (1951).

Chapter 4

RNA-DEPENDENT DNA POLYMERASES

Jerold A. Last and Theodore P. Zacharia

Harvard University
Cambridge, Massachusetts
and
National Academy of Sciences
Washington, D.C.

I.   INTRODUCTION

In 1970, Temin and Mizutani [15] and Baltimore [1] reported separately the discovery of RNA-dependent DNA polymerase ("reverse transcriptase"), which used endogenous RNA (viral or cellular) as a template for the synthesis of a complementary strand of DNA. Since then, the enzyme has been discovered in most of the known oncogenic RNA viruses, several slow-growing viruses, infected cells that produce viruses, infected cells that do not seem to produce viruses, and in uninfected cells. Thus, either

uninfected, infected, or malignant cells can produce double-
stranded DNA by the use of DNA-RNA hybrids as templates, or, in
the case of RNA viruses, the enzyme uses single-stranded RNA as
a template for the production of a DNA-RNA hybrid.

In this chapter, sources of enzyme, preparation of viruses,
and purification and assay of RNA-dependent DNA polymerases will
be described. Some terminology that will aid the less-sophisti-
cated reader to wend his way through a very jargon-enriched literature follows. A template is, in this field, a DNA or RNA
strand that directs the synthesis of a new polynucleotide strand,
whose sequence is complementary to that of the template itself.
A primer is a short oligonucleotide strand, complementary to a
short portion of the template, that can form a short, double-
helical region on the template, usually at an end, that represents the site of initiation of synthesis of the complementary
strand. Complementary implies the ability to form a base-paired
(helical) structure, e.g., adenine will base-pair with thymine
or uracil, and guanine is complementary with cytidine. Poly-
nucleotides are composed of several hundred or more nucleotides
and have molecular weights of hundreds of thousands to tens of
millions. Oligonucleotides are composed of several to several
tens of nucleotides and have molecular weights of thousands;
Oligo(dT)$_{14}$, for example, is a linear sequence of fourteen
deoxythymidylate residues (strictly speaking) or a population
of oligonucleotides of deoxythymidylate [also called $(dT)_{14}$]
whose average chain length is fourteen.

II. SOURCES OF POLYMERASES

Several of the sources of RNA-dependent DNA polymerases
that have been reported in the literature are listed in Table 1.
As is apparent, the most common sources are the tumor viruses
that contain RNA (rather than DNA) as their genetic material
(called the "viral genome"). While at one time the possession
of such an enzyme was taken to be a unique property of tumor
viruses (or tumor tissue), there have been several reports of
viruses not known to elicit tumors that contain the enzyme
(e.g., visna virus, syncytium-forming virus). There have also
been reports of apparently noncancerous tissue that contains
this enzyme (e.g., human embryonic tissue, normal lymphocytes).
Details of the occurrence of these enzymes in cells of specific
tissues are discussed under Preparation of Enzymes and Assays.

III. PURIFICATION OF VIRUSES

Rauscher murine leukemia virus and other oncogenic RNA
viruses can be obtained from plasma or from infected tissue
culture cells [14].

Plasma was clarified at 16,000 x g for 10 min. The supernatant was layered on a 100% glycerol cushion and centrifuged
at 95,000 x g for 70 min. The material above the glycerol
cushion was layered over a 25-50% sucrose gradient and centrifuged for 3 hr at 95,000 x g. Virus banded at a density of
1.16 g/cm^3. The virus was diluted in 10 mM Tris·HCl (pH 8.3)--
100 mM NaCl--2mM EDTA, and centrifuged again for 2 hr at 95,000 x g.

TABLE 1

Sources of RNA-Dependent DNA Polymerases

Source	Reference
Oncogenic viruses:	
Rous sarcoma	1, 15, 14
Human breast milk	7b
Rauscher murine leukemia	13, 2, 10b, 12, 14, 1, 6
Avian leukosis	4b
Avian myeloblastosis	13, 2, 7, 14
Avian reticuloendotheliosis	10c
Feline leukemia	13, 14
Feline sarcoma	4b
Moloney sarcoma	13, 14
Hamster leukemia	10c
Rat mammary tumor	13
Hamster sarcoma	10c
Mason-Pfizer monkey tumor	11, 14
Viper	10c
Lymphoid leukosis	14b
Mouse mammary tumor	13, 14
Slow-growing viruses:	
Primate syncytium-forming (foamy)	9b
Visna	10c

TABLE 1
(continued)

Sources of RNA-Dependent DNA Polymerases

Source	Reference
Mammalian cells	
Balb/3T3	12, 10b
Green monkey kidney	12
Leukemia lymphoblasts	5, 12
Wooley monkey	4b
HeLa	4
Gibbon ape	4b
Normal lymphocytes	10
Human embryonic lung	4
Human rhabdomyosarcoma RD-114	4b
Human skin fibroblasts	12
Leukemic mouse	3
Leukemic human plasma	8

The resultant pellet was suspended in the Tris--NaCl--EDTA buffer. The protein content was analyzed by the Lowry procedure [9a]. A similar procedure was used for the purification of Rauscher leukemia virus (RLV) that was grown and harvested from supernatants of tissue culture cells (JLS-V5).

Moloney sarcoma virus (MLV) was isolated from tumors that were induced in mice inoculated with the virus. Virus samples

were frozen and thawed twice, and then centrifuged for 10 min
at 16,000 x g. The supernatant was layered on a 20% glycerol
cushion on a 100% glycerol cushion, then centrifuged for 1 hr
at 95,000 x g. The material on the 100% glycerol bed was layered on a 25-50% sucrose gradient. Then, the procedure described
for RLV was used.

Mammary tumor virus was purified from fresh milk as described for RLV.

Feline leukemia virus can be obtained from feline cells
grown in suspension culture. The cell suspension was centrifuged for 10 min at 6400 x g; the supernatant was then clarified
for 10 min at 16,000 x g. Virus was precipitated by treatment
with 70% $(NH_4)_2SO_4$ and centrifuged for 10 min at 16,000 x g.
The pellet was suspended in Tris--NaCl--EDTA. The virus was
purified as described for MSV.

Avian myeloblastosis virus (AMV) was obtained from the
blood of chickens that were in the terminal stage of myeloblastic
leukemia, and was purified by the same methods as described
above.

Monkey mammary tumor virus was obtained from supernatant
fluids of monkey kidney cells grown in monolayers. The virus
was purified by two bandings in a zonal ultracentrifuge.

Rous sarcoma virus (RSV)--Rous associated virus-1 (RAV-1)
was obtained from cultures of chicken embryo fibroblasts. The
virus was purified as described for RLV.

Any of these purified viruses may be used as a crude preparation of RNA-dependent DNA polymerase, if the enzymatic activity is unmasked by the addition of a suitable nonionic detergent [e.g., Nonidet P-40(NP-40), Triton X-100, Lubrol, etc., see Assays].

IV. PURIFICATION OF POLYMERASES

Fridlender et al. [4] describe the preparation of two different peaks of activity (from a DEAE-cellulose column) of RNA-dependent DNA polymerase from HeLa cells as follows: Frozen cells were suspended in 20 ml of 10 mM NaCl--1 mM KPO_4 (pH 6.8) per gram of cells. The suspension was homogenized with several strokes of a Dounce homogenizer (an all-glass apparatus with a tight-fitting glass plunger in a glass cylinder). The nuclei were separated from the cytoplasm by centrifugation at 1,000 x g for 10 min. The nuclear pellet was resuspended in 5 volumes of 0.32 M sucrose--1 mM $MgCl_2$--1 mM KPO_4 (pH 7.0)--0.3% Triton detergent, clarified by centrifugation at 1,000 x g for 10 min, then centrifuged at 10,000 x g for 10 min. Under these conditions, about 90% of the total activity was solubilized. The soluble fraction was applied (1-4 mg of protein per g of DEAE-cellulose) to a column of DEAE-cellulose that had been equilibrated with 0.02 M KPO_4 (pH 7.5)--0.5 M dithiothreitol. The column was washed with 1 (column) volume of buffer, then eluted with 12 column volumes of a linear gradient from 0.02 M to 0.60 M KPO_4 in 0.5 M dithiothreitol. Two peaks of RNA-dependent DNA polymerase [as assayed with $(rA)_n \cdot (dT)_{12}$, see Section V] were eluted, at 0.10 and 0.12 M salt; while the two peaks were clearly resolved, no functional difference could be ascribed to the different fractions. Each of the peaks was separately pooled and added (3-5 mg of protein per g of phosphocellulose) to a (separate) column of phosphocellulose that had been equilibrated with 0.02 M KPO_4 (pH 8.9)--0.5 mM dithiothreitol. The column was washed with 1 (column) volume of the same buffer, and eluted with 10 column volumes of a linear gradient from 0.02 to 0.60 M KPO_4 (pH 8.9) in 0.5 M dithiothreitol. (See also [10b] for chromatography of RNA-dependent DNA polymerases on phosphocellulose.)

The recovery of enzyme(s) by this procedure is 60% (distributed 30% and 30% in the two peaks), with a total purification of 300-to 600-fold. The enzyme(s) may be clearly distinguished from, and separated from, the DNA-dependent DNA polymerase(s) from these cells by their chromatographic behavior, different optimal conditions for assay (see Section V), and different temperature optima (30°C rather than 37°C for the RNA-dependent, as compared with the DNA-dependent, polymerases).

Chargaff and his colleagues [3b] have purified more than 900-fold from chicken embryo a DNA polymerase (to greater than 90% purity) that prefers a DNA-RNA hybrid as its template (primer). The enzyme required Mn^{2+} as a cofactor and shows optimal activity in the presence of 120 mM KCl and a sulfhydryl-group reagent (e.g., dithiothreitol). The purification procedure

involves the following: embryonated eggs are candled, and the
embryos are removed and suspended in 0.35 M sucrose. A cell-free
extract is prepared in 0.35 M sucrose--50 mM Tris·HCl, pH 7.5--
25 mM KCl--10 $MgCl_2$ with a Potter-Elvejhem-type homogenizer
(glass barrel, Teflon plunger), and the material that is soluble
after centrifugation at 15,000 x g for 1 hr is retained. 2--
mercaptoethanol (to a final concentration of 5 mM) is added,
and proteins are "salted-out" of the extract by addition of
0.1 volume of 2 M acetate buffer. The precipitate is collected
at 10,000 x g and suspended in tris-acetate buffer. The enzyme
is then precipitated with between 40 and 80% of saturation of
ammonium sulfate after the removal of some extraneous proteins
by precipitation with: (i) 18%-saturated $(NH_4)_2SO_4$ and
(ii) 40%-saturated $(NH_4)_2SO_4$. The precipitated enzyme fraction
is dialyzed to remove $(NH_4)_2SO_4$ after solution in 50 mM Tris·HCl,
pH 7.5--0.35 M sucrose--0.3 M KCl--10 mM $MgCl_2$, and passed
through a column of DEAE-cellulose after dilution to lower the
salt concentration. About 30% of the total protein and most
of the nucleic acids in the extract stick to the column; the
enzyme passes through unadsorbed. Though purification is slight,
this is a rapid step and removes nucleic acids that would subsequently interfere with purification. The DEAE-cellulose eluate
was loaded on a column of carboxymethyl (CM)--Sephadex. The
column was washed, then eluted with 50 mM KPO_4, pH 6.5--0.4 M
KCl--5 mM 2-mercaptoethanol. Active fractions were pooled, precipitated with $(NH_4)_2SO_4$, and passed through a column of

CM-Sephadex; they could be further concentrated by dialysis

against 50% (v/v) glycerol in phosphate buffer. The active

concentrated enzyme was stored at -20°C under argon gas.

In an attempt to circumvent the problems of purification

of enzyme from a very limited amount of isolated virus, Yang

and coworkers [16] have used spleens from mice infected with

Rauscher leukemia virus as a starting material. Spleens were

isolated 3-4 weeks after infection, minced, and homogenized

in a glass-glass homogenizer in 4 volumes (8 ml per 2-gram

spleen) of 0.4 M sucrose, 50 mM Tris·HCl(pH 8.0), 50 mM KCl,

4 mM Mg(OAc)$_2$, and 2 mM dithiothreitol. The homogenate was

centrifuged at 1,000 x g for 10 min (pellet called nuclear

fraction), at 8,500 x g for 20 min (pellet called mitochondrial

fraction), and at 164,000 x g for 90 min (pellet called micro-

somal fraction). While RNA-dependent DNA polymerase activity

was present in several fractions, the microsomal pellet was the

source for further purification. Ribosome-bound cellular DNA

polymerase was removed from the pellet by elution of the enzyme

with 0.25 M KCl in 15% glycerol, 50 mM Tris·HCl(pH 8.0), 4 mM

Mg(OAc)$_2$, and 2 mM dithiothreitol. The presumptive viral poly-

merase was solubilized from washed microsomal fraction by

incubation for 10 min at 37°C with 0.5 M KCl-0.5% Nonidet P40

in the same glycerol-containing buffer. Particulate material

was removed by centrifugation for 90 min at 164,000 x g. The

supernatant, containing the RNA-dependent DNA polymerase, could

be stored at -70°C for at least a month without appreciable
loss of activity.

The enzyme could be purified about 800- to 1400-fold with
respect to the microsomal extract by a combination of hydroxyapatite chromatography, Sephadex G-100 filtration, and phosphocellulose chromatography. (Only details of the hydroxyapatite
step are given in ref 16; other steps are not described in detail.)

An elegant procedure for purification of the RNA-dependent
polymerase from leukemia and sarcoma viruses on a specific
solid-phase immunoadsorbent has been reported [9]. An antiserum
was prepared in rabbits against the specific polymerase. Immunoglobulin G was purified from the antiserum and was then covalently coupled to Sepharose 412, a dextran polymer. Crude
extracts of the viruses, which had been purified by sucrose
density gradient centrifugation, could be passed through this
immunoadsorbent column, which specifically retained a fraction
greatly enriched for the polymerase. The polymerase-containing
fraction could be eluted from the column by solvents of high pH
(alkaline) after removal of detergent (Triton X-100) by repeated
washes. The detergent must be removed for the recovery of
active enzyme.

Another technique for purification of the polymerase,
where the appropriate equipment is available, is electrofocusing--
separation of proteins on the basis of small differences in
their isoelectric points on a liquid column containing a gradient
of pH values that is stabilized within a sucrose density gradient.

Sonicated cell suspensions from normal human lymphocytes that
had been stimulated with phytohemagglutinin--a plant polysaccharide that stimulates lymphocytes to undergo mitosis and to
synthesize DNA--were purified [10] by the addition at 0°C of
2 ml of the lysate to a 110-ml column of ampholytes (the compounds responsible for the formation of a pH gradient; for
further information, see the technical literature furnished by
LKB Products, Uppsala, Sweden) that ranged in pH from 3 to 6,
in a 0 to 45% sucrose gradient. This column was subjected to
an applied (electric) voltage. The RNA-dependent enzyme could
be completely separated from the DNA-dependent polymerase by
this technique.

Gel-filtration chromatography of the polymerases from
murine leukemia virus and its host cells (cultured mouse fibroblasts) on Sephadex G-100 has been reported [10b], and is another
potentially useful technique for purification of enzymes from
crude extracts.

V. ASSAYS

The common components of a typical assay mixture are:
tris buffer, at a pH of about 7.5 to 8.5 and a concentration
around 50 mM; a divalent cation, either Mg^{2+} or Mn^{2+}, 6-8 mM;
a monovalent cation, often K^+ or Na^+, with a wide range of
concentrations from about 40 to 125 mM; an SH-group reductant,
such as dithiothreitol, may be added at 2-100 mM; a labeled
deoxyribonucleotide triphosphate (either ^3H, ^{14}C, or α-^{32}P is

appropriate); a template polyribonucleotide complementary to
the labeled deoxyribonucleotide triphosphate used [e.g., poly(rA)
for dTTP], a deoxyribooligonucleotide primer complementary to
the template, such as oligo (dT) for poly (rA), with an average
length of about 8-20 residues of dT; and 0.05--0.2% of a
nonionic detergent to disrupt virion structure sufficiently
to expose the enzyme when purified (or crude) virus particles
are assayed--the detergent is omitted when purified, soluble
enzymes are tested. The specificity of the assay with poly(rA)·
oligo(dT) as primer-template for distinguishing viral and cellular DNA polymerases has been challenged [10a].

Baltimore and Smoler [2] assayed the polymerase from
avian myeloblastosis virus in the following mixture: 50 mM
Tris·HCl (pH 8.3), 6 mM Mg(acetate)$_2$, 20 mM dithiothreitol,
60 mM NaCl, 0.2% NP-40 (Nonidet P-40, Shell Chemical Co.),
1100-1650 pmole of poly(U) [or poly(C), or poly(I)], 94-274
pmole of dATP [or dGTP, as appropriate], and 200-300 pmole of
d(T)$_{14}$ [or (dG)$_{14}$, as appropriate]. Reactions were incubated
at 37°C for 60 min, and radioactivity incorporated into acid-
precipitable material was determined.

The polymerase from murine leukemia virus was assayed in
exactly the same way, except 1 mM MnCl$_2$ was used in place of
6 mM Mg(acetate)$_2$, and the NP-40 concentration was 0.05% [2].

Under these conditions, one complete strand of deoxyribonucleotide, complementary to the template strand, was
synthesized before incorporation stopped.

Spiegelman's group [11] assayed the polymerase from the
Mason-Pfizer monkey tumor virus in a 250-µl reaction mixture
of: 48 mM Tris·HCl (pH 8.3), 6.8 mM $MgCl_2$, 40 mM KCl, 0.8
mM (each) dATP, dGTP, DCTP, 0.04 M [^{3}H]dTTP (1700 cpm/pmole),
and 250 µg of purified virus (previously incubated at 0°C for
10 min with 0.2% NP-40 and 0.1 M dithiothreitol). This mixture
was incubated 60 min at 37°C, then precipitated with trichloroacetic acid (TCA) containing inorganic phosphate and inorganic
pyrophosphate [11] and added RNA from E. coli as a carrier.
The insoluble material was collected on a Millipore filter,
washed, dried, and counted in a liquid scintillation counter
in a suitable scintillation counter in a suitable scintillation
solvent.

Weissbach and co-workers [4] assayed purified RNA polymerases from tissue-culture cells in the following 200-µl mixture: 50 mM Tris·HCl (pH 7.5), 0.5 mM $MnCl_2$, 2.5 mM dithiothreitol, 7.5 µg of $(dT)_{12}·(rA)_n$, 20 nmole of [^{3}H]dTTP, 125 mM
KCl, 90 µg of bovine serum albumin [added to stabilize the
enzyme protein], and 10-50 µl of enzyme solution. Tubes were
incubated for 15 min at 30°C and cooled to 0°C; the product
was precipitated with TCA, collected, washed, and counted.
Several points are noteworthy about this particular assay for
RNA-dependent DNA polymerases from HeLa cells (and other tissue-
culture lines). Note the use of Mn^{2+} for Mg^{2+} (Mg^{2+} only

weakly replaced Mn^{2+}; there is said to be 20% of the activity at 5 mM $MgCl_2$, ten times the optimal Mn concentration), the use of 30°C (the enzyme is only half as active at 37°C), the preference for a poly(ribonucleotide) template, and the addition of a high concentration of KCl for optimal activity. The DNA-dependent DNA polymerases from such tissue culture cells are optimally assayed without KCl; with Mg^{2+}, not Mn^{2+}; at 37°C; and with a poly(deoxyribonucleotide) template.

ACKNOWLEDGMENT

We acknowledge Dr. W. Parks and Dr. E. Scolnick, without whose inspiration we would have never considered writing this paper.

REFERENCES

[1] D. Baltimore, Nature, 226, 1209-1211 (1970).

[2] D. Baltimore and D. Smoler, Proc. Nat. Acad. Sci. USA, 68, 1507-1511 (1971).

[3] H. B. Bosmann, FEBS Lett., 13, 121-123 (1971).

[3a] J. G. Stavrianopoulos, J. D. Karkas, and E. Chargaff, Proc. Nat. Acad. Sci. USA, 69, 1781-1785 (1972).

[4] B. Fridlender, M. Fry, A. Bolden, and A. Weissbach, Proc. Nat. Acad. Sci. USA, 69, 452-455 (1972).

[4a] R. C. Gallo, personal communication.

[5] R. C. Gallo, S. S. Yang, and R. C. Ting, Nature, 228, 927-929 (1970).

[6] B. I. Gerwin, G. J. Todaro, V. Zeve, E. M. Scolnick, and

S. A. Aaronson, Nature, 228, 435-438 (1970).

[7] N. C. Goodman and S. Spiegelman, Proc. Nat. Acad. Sci. USA,

68, 2203-2206 (1971).

[7b] M. Green, Proc. Nat. Acad. Sci. USA, 69, 1036-1041 (1972).

[8] A. A. Kiessling, G. H. Weber, A. O. Delney, E. A. Possehl,

and G. S. Beaudreau, J. Virol., 7, 221-226 (1971).

[9] D. M. Livingston, E. M. Scolnick, W. P. Parks, and G. J.

Todaro, Proc. Nat. Acad. Sci. USA, 69, 393-397 (1972).

[9a] O. H. Lowry, N. J. Rosebrough, A. L. Farr, and R. J.

Randall, J. Biol. Chem., 193, 265-275 (1951).

[9b] W. P. Parks, E. M. Scolnick, G. J. Todaro, and S. A.

Aaronson, Nature, 229, 259-261 (1971).

[10] P. E. Penner, L. H. Cohen, and L. A. Loeb, Biochem.

Biophys. Res. Commun., 43, 1228-1234 (1971).

[10a] M. S. Robert, R. G. Smith, R. C. Gallo, P. S. Sarin,

and J. W. Abrell, Science, 176, 798-800 (1972).

[10b] J. Ross, E. M. Scolnick, G. J. Todaro, and S. A.

Aaronson, Nature, New Biol., 231, 163-167 (1971).

[10c] J. Schlom, D. H. Harter, A. Burny, and S. Spiegelman,

Proc. Nat. Acad. Sci. USA, 68, 182-1   (1971).

[11] J. Schlom and S. Spiegelman, Proc. Nat. Acad. Sci. USA,

68, 1613-1617 (1971).

[12] E. M. Scolnick, S. A. Aaronson, G. J. Todaro, and W. P.

Parks, Nature, 229, 318-321 (1971).

[13] S. Spiegelman, A. Burny, M. R. Das, J. Keyder, J. Schlom, M. Travnicek, and K. Watson, Nature, 228, 430-432 (1970a).

[14] S. Spiegelman, A. Burny, M. R. Das, J. Keydar, J. Schlom, M. Travnicek, and K. Watson, Nature, 227, 563-567 (1970b).

[14b] H. M. Temin, Proc. Nat. Acad. Sci. USA, 69, 1016-1020 (1972).

[15] H. M. Temin and S. Mizutani, Nature, 226, 1211-1213 (1970).

[16] W. K. Yang, C. K. Koh, and L. C. Waters, Biochem. Biophys. Res. Commun., 47, 505-511 (1972).

Chapter 5

RIBONUCLEASE H OF <u>ESCHERICHIA</u> <u>COLI</u>

Ira Berkower

Department of Developmental Biology and Cancer
Albert Einstein College of Medicine
Bronx, New York

I. INTRODUCTION

A ribonuclease activity that specifically degrades RNA in RNA·DNA hybrid structures was first described by Hausen and Stein [1] in calf thymus. This enzyme, called RNase H, catalyzed the hydrolysis of [^{3}H]-poly U·poly dA, but not [^{3}H]-poly U·poly A or ribosomal RNA. In addition, RNA·DNA hybrids generated by the action of RNA polymerase on single-stranded DNA templates were degraded before, but not after, heat denaturation. RNase H activity has been detected from a variety of sources, including human KB cells [2], RNA tumor virions [3-5] (associated

145

with reverse transcriptase), and Escherichia coli [6-9]. The
current hypothesis that short RNA segments serve as primer
strands for initiation of DNA replication in M13 phage [10] and
in Okazaki pieces [11,12] suggests a biological role for RNase
H of removing these primers and allowing the sealing of continuous DNA by DNA ligase.

II. ASSAY

RNase H activity is assayed by the conversion of acid-
insoluble ribohomopolymers to acid-soluble oligonucleotides,
dependent on the addition of the complementary deoxyhomopolymer.
Reaction mixtures (0.05 ml) contain 1.2 nmole [^{3}H]-poly A (32
cpm/pmole, 400 nucleotides in length, labeled ribohomopolymers
available from Schwarz/Mann Bioresearch), 1.2 nmole poly dT
(deoxyhomopolymers available from General Biochemicals), 2 μmole
tris·HCl (pH 7.7), 0.2 mole MgCl$_2$, 0.05 μmole dithioerythritol,
1.5 μg bovine serum albumin, 0.005 ml 40% glycerol, and varying
amounts of RNase H. After 20 min incubation at 30°C, the reaction is stopped by adding 0.1 ml of cold 0.1 M sodium pyrophosphate, 0.05 ml of denatured salmon-sperm DNA (1 mg/ml), 0.1 ml
albumin (10 mg/ml), and 0.3 ml of 10% trichloroacetic acid.
The acid-insoluble material is removed by centrifugation for
2 min at 4000 rpm in an International refrigerated centrifuge,
and the supernatant fluid is counted in 10 ml of Bray's fluid
[13] in a scintillation counter. One unit of activity is defined as the amount of enzyme producing 1 nmole of acid-soluble

material in 20 min at 30°C. The assay of poly A degradation
is proportional to enzyme concentration between 0.05 to 0.5
unit. The reaction is linear for more than 60 min and up to
60% degradation of poly A. However, the poly A·poly dT assay
is unreliable for the assay of crude fractions, which contain
RNases capable of acting on poly A in the absence of poly dT.
Instead, annealed [^{3}H]-poly I·poly dC mixtures, which are resistant to RNase T$_2$, should be used. Since commercially available [^{3}H]-poly I gives high background levels (10%) of acid-
soluble oligonucleotides, the assay is performed as described
above, except that the acid-precipitable material is collected
on Gelman type E glass fiber filters, and its radioactivity
measured by scintillation counting. Enzymatic activity is
quantitated by the decrease in acid-precipitable material.

III. PURIFICATION

All operations are carried out at 4°C. Two hundred grams
of commercially grown E. coli B are ground to a thick paste
with 600 gm of glass beads in a Waring Blendor in 500 ml of a
solution containing 5 M NaCl, 0.05 M tris·HCl (pH 8), 0.1 mM
dithiothreitol, and 0.1 mM EDTA (buffer A). Glass beads are
removed by centrifuging 20 min at 5000 x g. DNA is removed by
PEG/Dex phase partition [14] as follows: To each 100 ml of crude
extract are added 32.1 ml of 30% polyethylene glycol and 11.2 ml
of 20% dextran T 500; the mixture is stirred 30 min, then centrifuged 10 min at 1,000 x g to produce two distinct phases.

The upper phase (1300 ml free of DNA) is dialyzed overnight against buffer A without NaCl and then passed through a 300 ml DNA cellulose column (containing 0.5 mg denatured calf thymus DNA per ml packed volume). The column is washed with 0.15 M NaCl in buffer A until the OD_{280} of the eluate is less than 0.08. RNase H activity is eluted with 200 ml of 1.2 M NaCl in buffer A, along with RNA polymerase, DNA polymerases, and other DNA binding proteins [15].

The eluate fraction is concentrated by precipitation with 1.5 volumes of saturated, neutralized ammonium sulfate and is resuspended in approximately 10 ml of 0.02 M tris·HCl (pH 8), 0.05 M NaCl, 0.1 mM dithiothreitol, and 0.1 mM EDTA in 5% glycerol (buffer B), and dialyzed 3 times against 500 ml buffer B. The dialysate is passed through a 250 ml 5 m agarose column (50 x 2.5 cm) equilibrated with buffer B. RNase H activity elutes after 1 column volume, and active fractions are pooled and dialyzed overnight against 1 liter of 0.01 M tris·HCl (pH 8), 0.01 M 2-mercaptoethanol, and 0.1 mM EDTA in 30% glycerol (buffer C). RNA polymerase elutes in earlier fractions (0.58 column volume) and can be purified further.

The dialyzed 5-m agarose fraction (salt conductivity 0.02 M NaCl, 13 mg protein) is applied to a DEAE -cellulose column (15 x 2.1 cm) equilibrated with buffer C in 10% glycerol. The column is eluted with a 400-ml linear gradient of NH_4Cl (0 to 0.6 M), and 6 ml fractions are collected. Two major peaks of

RNase H activity will be noted, eluting at 0.07 M (peak A) and
0.14 M NH$_4$Cl (peak B). These peaks are pooled separately and
dialyzed against 0.02 M potassium phosphate (pH 6.5), 0.01 M
2-mercaptethanol and 0.1 mM EDTA in 50% glycerol (buffer D).
Peak A contains 50% of the total RNase H activity applied to
the column, while peak B represents 20% of the original activity.
Peak A is inhibited more than 90% by 20 mM N-ethylmaleimide,
while peak B is resistant. In addition, peak B is sensitive to
antiserum prepared against purified DNA polymerase I and copurifies
with polymerizing activity during chromatography on phosphocellulose. This activity degrades hybrid RNA in the 5' to 3' direction and is tentatively identified as the 5' to 3' exonuclease
of DNA polymerase I [5, 16, 17].

   Peak A (0.2 mg protein) is applied to a phosphocellulose
column (10 x 2.1 cm) equilibrated with buffer D in 10% glycerol.
The enzyme is eluted with a 300-ml linear gradient of 0.02 to
0.6 M potassium phosphate (pH 6.5) in buffer D, and 5 ml fractions are collected. Ordinarily, a single peak of RNase H
activity is observed, eluting at 0.21 M potassium phosphate.
However, if peaks A and B were not well resolved by DEAE-
cellulose chromatography, they will be separated by the phosphocellulose step, giving a second peak at 0.1 M potassium phosphate,
which is resistant to N-ethyl maleimide and corresponds to DNA
polymerase I. Depending on the purity of the DEAE-cellulose
fraction, this step will give a 1-to-10-fold purification, with
complete recovery of activity.

The phosphocellulose fraction still contains a contaminating RNase activity that attacks poly U and poly C but not poly A or poly I. This contaminant is heat stable and Mg^{2+} requiring, and it would interfere with studies of the substrate specificity of RNase H. It is removed from the phosphocellulose fraction by chromatography on a Sephadex G-100 column (25 x 1.1 cm), equilibrated with 0.04 M potassium phosphate (pH 6.5), 0.01 M 2-mercaptoethanol, 0.1 mM EDTA in 20% glycerol. One ml fractions are collected and assayed for both RNase activity, using $[^3H]$-poly U, and RNase H activity, using $[^3H]$-poly A·poly dT. The pyrimidine-specific RNase elutes near the V_o of the Sephadex column, while RNase H is included in the gel. The enxyme is free of detectable DNase activity toward $[^3H]$-fd DNA when examined in neutral sucrose gradients. However, unpublished results of Dr. Larry Grossman show that some RNase H preparations were contaminated with endonuclease I, which converts double-stranded circular PM-2 DNA (form I) to the relaxed form II when centrifuged to equilibrum in cesium chloride density gradients containing ethidium bromide. Under conditions where this contaminating activity on double-stranded DNA is completely inhibited by the addition of tRNA, the RNase H activity retains the ability to convert closed-circular mitochondrial DNA Form I, containing covalent RNA, to Form II, presumably by specific nicking of the covalent RNA segment.

A summary of the purification of RNase H from E. coli B is presented in Table 1. The RNase H activity is purified approximately 2000-fold over crude extracts, with a yield of 4%. The phosphocellulose fraction can be stored in 0.05M tris·HCl, pH

7.7, 0.05 M NaCl, 0.01 M 2-mercaptoethanol, 0.1 mM EDTA, and

50% glycerol at -20°C, and is stable for more than 6 months.

Although no specific inhibitors are known, heating 5 min at

65°C or overnight dialysis against buffers containing no glyc-

erol are means of removing RNase H activity from preparations

of other enzymes. RNase H has also been purified from E. coli

strain D110 (<u>pol</u> A_1^-, <u>endo</u> I^-, <u>thy</u>$^-$), which provides enzyme free

of endonuclease I, and with properties identical to those des-

cribed below for the enzyme purified from <u>E. coli</u> B.

IV. REQUIREMENTS AND SPECIFICITY

For optimal activity the enzyme requires 2-4 mM Mg^{2+},

which can be replaced only minimally (<5% activity) by Mn^{2+}.

The optimum pH is 7.5 to 9.1, with half-maximal activity at

pH 6.9. Dithioerythritol (1 mM) or other sulfhydryl group is

important, and the activity is inhibited completely by N-ethyl

maleimide. RNase H is relatively insensitive to salt, retain-

ing 50% activity at 0.3 M NaCl.

The specificity of RNase H for various synthetic homopoly-

mers is shown in Table 2. Degradation of each labeled ribo-

homopolymer requires the complementary deoxyribohomopolymer,

which cannot be replaced by the complementary ribohomopolymer.

The rate of degradation varies as much as 10-fold, with poly I·

TABLE 1

Purification of RNase H from E. coli B

Fraction	Protein (mg)	Activity (units)	Specific Activity (units/mg)	Yield (%)
Crude extract	13,600	430,000	32	100
Polyethylene glycol phase	4,550	355,000	78	82
5 m agarose	12.9	34,000	2,640	8
DEAE-cellulose Peak A	0.2	15,600	78,000	3.6
Peak B	4.2	6,430	1,530	1.5
Phosphocellulose of Peak A	0.2	15,700	79,000	3.6

TABLE 2

Substrate Specificity of RNase H[a]

Additions	RNase H activity pmol/20 min/μg protein
[3H]-poly A	<1
+ poly dT	54
+ poly U	<1
[3H]-poly U	<0.1
+ poly dA	7
+ poly A	<0.1
[3H]-poly C	<0.5
+ poly dG	68
+ poly G	<0.2

TABLE 2
(continued)

Substrate Specificity of RNase H [a]

Additions	RNase H activity pmol/20 min/µg protein
[^{3}H]-poly I	<3
+ poly dC	260
+ poly C	<3
fd DNA-[^{3}H]RNA	15

[a] RNase H isolated from E. coli B was incubated with poly-nucleotides as described in the text. The amount of polynucleo-tides in each case was as follows: [^{3}H]-poly A 1.2 nmoles, poly dT 1.75 nmole, poly U 0.94 nmole, [^{3}H]-poly U 0.05 nmole, poly dA 1.35 nmole, poly A 1.0 nmole, [^{3}H]-poly C 2 nmole, poly dG 2.5 nmole, poly G 2 nmole, [^{3}H]-poly I 0.06 nmole, [^{3}H]-poly I·poly dC 0.22 nmole, [^{3}H]-poly I·poly C 0.06 nmole, and fd DNA·[^{3}H] RNA 0.008 nmole. In experiments with [^{3}H]-poly I and fd DNA·[^{3}H] RNA, acid-insoluble radioactivity was determ-ined.

poly dC and poly C·poly dG showing faster degradation than poly
A·poly dT and poly U·poly dA. The rate of degradation of RNA
with fd DNA·RNA hybrid structures (formed by RNA polymerase
acting on single-stranded fd DNA [18]) is in the same range of
activity as with synthetic polymers.

V. PRODUCTS AND MECHANISM OF ACTION

When the products of complete digestion of [^{3}H]-poly A·poly
dT by RNase H are chromatographed on DEAE-cellulose in 7 M urea,
they are resolved into oligonucleotide peaks composed of AMP 4%,
$(pA)_2$ 16%, $(pA)_3$ 31%, $(pA)_4$ 27%, $(pA)_5$ 19%, $(pA)_{6-8}$ 2.6%. Treatment of the $(pA)_5$ material with bacterial alkaline phosphatase
results in a decrease of negative charge and comigration with
standard $(Ap)_4A$ on electrophoresis at pH 3.5. Snake venom
phosphodiesterase [19] completely digests the pentanucleotide
(not treated with alkaline phosphatase) to AMP at pH 8.6. Thus,
the products contain 5'-phosphate and 3'-hydroxyl termini.

RNase H acts by endonucleolytic attack [20]. Poly A can be
labeled at the 5' end with [γ-^{32}P]-ATP and polynucleotide kinase
and extended at the 3' end with [^{3}H]-ADP and polynucleotide
phosphorylase. The label at both ends is rendered acid soluble
at an equal rate in the presence of poly dT. In addition, poly
A substrates, which are made circular by RNA ligase [21], or
which are blocked at one or both ends with covalently bound
cellulose or poly C or both, are all susceptible to RNase H
attack in the presence of poly dT. Since these modified poly A
substrates lack hybrid RNA ends, the enzyme must act endonucleo-
lytically.

REFERENCES

[1] P. Hausen and H. Stein, Eur. J. Biochem., 14, 278 (1970).

[2] W. Keller and R. Crouch, Proc. Nat. Acad. Sci. USA, 69, 3360 (1972).

[3] J. Leis, I. Berkower, and J. Hurwitz, in DNA Synthesis in Vitro (in Steenbock Symp.) (R. Wells and R. Inman, eds.), University Park Press, Maryland, in press.

[4] J. Leis and J. Hurwitz, J. Biol. Chem., in press, (1973).

[5] D. Baltimore and D. F. Smoler, J. Biol. Chem., 247, 7282 (1972).

[6] H. I. Miller, G. N. Gill, and A. D. Riggs, Fed. Proc., Fed. Amer. Soc. Exp. Biol, 31, 500 (1972).

[7] H. D. Robertson, R. E. Webster, and N. D. Zinder, J. Biol. Chem., 243, 82 (1968).

[8] H. D. Robertson, Nature New Biol., 229, 169 (1971).

[9] I. Berkower, J. Leis, and J. Hurwitz, J. Biol Chem., in press (1973).

[10] W. Wickner, D. Brutlag, R. Schekman, and A. Kornberg, Proc. Nat. Acad. Sci. USA, 69, 965 (1972).

[11] A. Sugino, S. Hirose, and R. Okazaki, Proc. Nat. Acad. Sci. USA, 69, 1863 (1972).

[12] A. Sugino and R. Okazaki, Proc. Nat. Acad. Sci. USA, 70, 88 (1973).

[13] G. A. Bray, Anal. Biochem., I, 279 (1960).

[14] B. M. Alberts, in Methods in Enzymology, (L. Grossman and
K. Moldave, eds.), Vol. XII, Part A., p. 566, Academic Press,
New York, 1967.

[15] H. Schaller, C. Nüsslein, F. J. Bonhoeffer, C. Kurz, and
I. Nietzschmann, Eur. J. Biochem., 26, 474 (1972).

[16] R. P. Klett, A. Cerami, and E. Reich, Proc. Nat. Acad.
Sci. USA, 60, 943 (1968).

[17] M. P. Deutscher and A. Kornberg, J. Biol. Chem., 244,
3029 (1969).

[18] U. Maitra and J. Hurwitz, Proc. Nat. Acad. Sci. USA, 54,
815 (1965).

[19] H. G. Khorana, in The Enzymes, 2nd Ed. Vol. 5, p. 86, (P. D.
Boyer, H. Lardy, and K. Myrback, eds.) Academic Press, New York,
1961.

[20] J. Leis, I. Berkower, and J. Hurwitz, Proc. Nat. Acad.
Sci. USA, 70, 466 (1973).

[21] R. Silber, V. G. Malathi, and J. Hurwitz, Proc. Nat. Acad.
Sci. USA, 69, 3009 (1972).

Chapter 6

DNA POLYMERASE II

Reed B. Wickner

Department of Developmental Biology and Cancer
Albert Einstein College of Medicine
Bronx, New York

I.   INTRODUCTION

DNA polymerase I, the enzyme studied extensively by Kornberg
and his collaborators [1] (the purification of DNA polymerase I
is described by Jovin et al. [2]) is not necessary for DNA replication, as shown by the discovery of a mutant (pol A_1), deficient in this enzyme, which grows normally [3]. DNA polymerase
I can, however, repair UV-damaged DNA both in vivo [4] and in
vitro [5]. The discovery of the Cairns mutant has both prompted
and encouraged the search for other DNA-synthesizing activities.

157

DNA polymerase II [6-11] has been purified extensively by several groups. The method discussed below is most coveniently carried out using the Cairns mutant or an endonuclease I (end$^-$)-deficient derivative thereof, such as E. coli D110 isolated by Moses and Richardson [12].

II. MATERIALS

Escherichia coli D110 (end$^-$, polA$_1$, thy$^-$) is a derivative of M3110 and was obtained from R. Moses.

LB phosphate medium contains, per liter, 10 gm trypticase, 5 gm NaCl, 5 gm yeast extract, 1 ml 1 N NaOH, 10 mg thiamine, 3.5 gm K_2HPO_4, and 1.5 gm KH_2PO_4.

Preparation of Partially Degraded Salmon-Sperm DNA: The reaction mixture (3ml) contains 18 μmole of salmon sperm DNA, 30 μmoles of tris chloride (pH 7.5), 15 μmole $MgCl_2$, and 0.5 μg pancreatic DNase (Worthington). After 30 min at 38°C, the mixture is heated to 65°C for 10 min to inactivate the DNase. Under these conditions, 20% of the DNA is rendered acid-soluble.

III. ASSAY OF DNA POLYMERASE II

The standard reaction (0.1 ml) contains 6.0 μmole tris chloride (pH 7.5), 0.84 μmole $MgCl_2$, 0.15 μmole dithiothreitol, 5 nmole each dATP, dGTP, and dCTP, 1 nmole [^{3}H]dTTP (400 cpm per pmole), and 22 nmole partially degraded salmon-sperm DNA. The reaction is initiated by the addition of enzyme, and after a 30-min incubation at 37°C the reaction is terminated by the addition of 0.1 ml 0.1M sodium pyrophosphate containing 0.5 mg/

ml of bovine serum albumin, followed by the addition of 3 ml of 5% trichloroacetic acid. The precipitate is collected on a glass fiber filter and washed with 20 ml of 1% trichloroacetic acid and 5 ml of ethanol. The filter is dried under a heat lamp and counted in a toluene-based scintillation fluid. One polymerase unit is defined as the amount of enzyme catalyzing the incorporation of 1 nmole of dTMP per 30 min at 37°C into DNA.

IV. PURIFICATION PROCEDURE

The procedure described below was developed in our laboratory [9-11]. Other procedures have been described by Moses and Richardson [7], Kornberg and Gefter [6], and Knippers [8].

Step 1: Crude Extract

Escherichia coli D110 (pol A$_1$, end$^-$) is grown at 37°C with vigorous aeration in a 100-liter fermenter to OD$_{650}$ = 1.0 in LB phosphate medium containing 10 µg/ml thymine. However, the specific activity of DNA polymerase II in crude extracts is essentially independent of the growth phase in which the cells are harvested. The cells are collected by centrifugation in a Sharples centrifuge and frozen at -20°C until used. All subsequent steps are carried out at 0°-5°C. Partially thawed cells (178 gm) are ground with 356 gm alumina and the mixture is suspended in 534 ml of 0.05 M tris chloride (pH 7.5) containing 1 mM dithiothreitol and 0.1 mM EDTA. This suspension is centrifuged at 10,000 x g for 10 min, the precipitate is discarded, and the supernatant is recentrifuged at 100,000 x g for 90 min. The supernatant

fluid is decanted and used for subsequent steps. The assay at
this point is linear only at low levels of enzyme.

Step 2: Polyethylene Glycol-Dextran Phase Partition

To the supernatant from step 1 (500 ml) are added 7.85 gm
dextran 500 T, 33.7 gm polyethylene glycol 6000, and 135 gm NaCl.
This mixture is stirred for 30 min and the phases are then
separated by centrifugation at 16,000 x g for 10 min. The
lower phase is discarded and the upper phase dialyzed three
times for 8 hr each, against 8 liter of 0.02 M tris chloride,
pH 7.5, containing 1 mM dithiothreitol and 0.1 mM EDTA.

Step 3: DEAE-Cellulose Chromatography

The product of step 2 is applied to a 4.2 cm x 29 cm
column of DEAE-cellulose (DE-52, Whatman) previously equilibrated with 0.02 M tris chloride, pH 7.5, containing 1 mM
2-mercaptoethanol and 0.1 mM EDTA. The column then is washed
with 200 ml of this buffer and developed with a linear gradient
of 3 liters of 0.05-0.25 M KCl in the same buffer. DNA polymerase
II is eluted between 0.12 and 0.18 M KCl. Some of these fractions contain exonuclease III, while exonuclease I is eluted
well after DNA polymerase II. Because large amounts of exonuclease III interfere with the assay by degrading the product,
there often appear to be two peaks of polymerase activity
eluting from the DEAE-cellulose column. Both peaks and the
valley between them must be pooled to avoid great losses of
enzyme at this step. The pooled fractions are dialyzed twice

for 4 hr each time against 4 liter 0.02 M potassium phosphate
buffer (pH 6.5), containing 1 mM 2-mercaptoethanol, 0.1 mM
EDTA, and 10% v/v glycerol.

Step 4: First Phosphocellulose Column Chromatography

Phosphocellulose to be used for column chromatography
(P-11,Whatman) is washed by decantation first with 0.1 N HCl
in 50% ethanol, then with water, followed by 0.1 N NaOH, and
finally with water before use. The dialyzed product of step 3
is loaded and washed onto a 4.2 cm x 29 cm phosphocellulose
column equilibrated with the buffer used for dialysis in step
3. The column is then washed with a linear gradient of 3 liter
potassium phosphate buffer (pH 6.5), from 0.05 to 0.6 M containing 1 mM 2-mercaptoethanol, 0.1 mM EDTA, and 10% glycerol; the
enzyme is eluted between 0.2 and 0.3 M potassium phosphate.
The specific activity and apparent yield are reduced due to the
presence of exonuclease III in the same fractions.

Step 5: Second Phosphocellulose Column Chromatography

The active fractions isolated in step 4 are pooled, again
dialyzed against 0.02 M potassium phosphate (pH 6.5), containing 1 mM 2-mercaptoethanol, 0.1 mM EDTA, and 10% glycerol,
loaded onto a column identical to that used in step 4, and
washed onto the column with 200 ml of the same buffer. The
column is then developed with a linear gradient of 3 liter 0.05-
0.6 M KCl in the same buffer used in step 4. DNA polymerase II
is eluted at about 0.5 M KCl, now well separated from exonuclease III, which elutes much earlier.

The peak fractions are pooled and concentrated 20-fold by
dialysis against 0.05 M tris chloride (pH 7.5), 1 mM 2-mercap-
toethanol, 0.1 mM EDTA, and 10% glycerol containing 30% polyethylene glycol. The concentrated enzyme is made 30% with
glycerol and stored at 0°C. Under these conditions the enzyme
is completely stable for at least 8 months. The enzyme is inactivated by freezing and thawing. A summary of the purification is presented in Table 1.

To remove traces of RNase in this preparation, the enzyme
may be further purified by gel filtration on Sephadex G-150.

Purity of the Final Product

The final product of the purification is free of RNase,
DNA ligase, endonuclease activity on single- or double-stranded
DNA, and exonucleolytic activity on double-stranded DNA. The
purified enzyme is probably not homogeneous [10]; indeed,
there is as yet no published evidence for the purification of
DNA polymerase II to homogeneity. The purified enzyme possesses
an exonucleolytic activity degrading single-stranded DNA in the
3'- to 5'-direction [6, 13, 14].

This activity probably resides in the same molecule as
the polymerase activity.

V.   PROPERTIES OF DNA POLYMERASE II

The enzyme requires all four deoxynucleoside triphosphates
and Mg^{2+} and is unaffected by added ATP [6-10]. The DNA tem-

TABLE 1

Purification of DNA Polymerase II

Step	Protein (mg)	Activity (units)	Specific activity (units/mg)	Purification (-fold)	Yield (%)
1. Crude extract[a]	11,000	1,100	0.1	(1)	(100)
2. Polyethylene glycol-dextran phase partition	3,500	880	0.25	2.5	80
3. DEAE-cellulose	500	710	1.3	13	64
4. Phosphocellulose (PO$_4$ elution)	12.3	200[b]	16.2[b]	162[b]	18[b]
5. Phosphocellulose (KCl elution)	0.25	700	2,660	26,600	64

[a]The starting material was 178 gm (wet weight) of E. coli D110.

[b]The apparent low yield and poor purification at this step is due to the coincident elution of exonuclease III with the peak of DNA polymerase II.

plate must contain a 3'-hydroxyl primer strand and an adjacent
single-stranded template strand of less than about 100 nucleotides in length [11]. The enzyme carries out nucleotide incorporation in the 5' to 3' direction, like all other described
DNA polymerases, and has a very low error frequency [14]. DNA
polymerase II can completely and exactly convert the single-
stranded region of the template DNA into a duplex structure,
whether the single-stranded template region was internal (a gap)
or at the end of a duplex structure. Unlike DNA polymerase I,
however, it will stop at this point. DNA polymerase II is
unable to initiate synthesis at a single-stranded scission in
native DNA and thus can neither displace nor degrade the 5' end
of the adjacent strand. The product of synthesis is covalently
attached to the primer strand [8, 11, 14].

The enzyme is present in the small, DNA-negative "minicells."
Thus, DNA polymerase II is not in constant association with the
bacterial chromosome [11].

In summary, DNA polymerase II carries out only "repair"
synthesis in vitro; that is, it does not initiate new DNA
chains. It is present in crude extracts at a level sufficient
only to account for 0.3% of the in vivo rate of DNA synthesis.
Added factors may, however, allow the enzyme to carry out
processes it cannot catalyze alone and do these at a faster
rate.

Recently, mutants, the extracts of which are deficient in

DNA polymerase II, have been isolated [15,16]. These strains grow normally and support the growth of bacteriophages T4, T7, λ, ØX174, fd, P2, and many others. They show normal repair of DNA after UV irradiation or treatment with alkylating agents, and show normal recombination frequencies. Thus, the role of DNA polymerase II *in vivo* remains unclear.

REFERENCES

[1] A. Kornberg, Science, 163, 1410 (1969).

[2] T. M. Jovin, P. T. Englund, and L. L. Bertsch, J. Biol. Chem., 244, 2996 (1969).

[3] P. DeLucia and J. Cairns, Nature (London), 224, 1164 (1969).

[4] J. D. Gross and M. Gross, Nature (London), 224, 1166 (1969).

[5] R. B. Kelly, M. R. Atkinson, J. A. Huberman, and A. Kornberg, Nature (London), 224, 495 (1969).

[6] T. Kornberg and M. Gefter, Biochem. Biophys. Res. Commun., 40, 1348 (1970); Proc. Nat. Acad. Sci. USA, 68, 761 (1971).

[7] R. Moses and C. C. Richardson, Biochem. Biophys. Res. Commun., 41, 1557; 1565, (1970).

[8] R. Knippers, Nature (London), 228, 1050 (1970).

[9] B. Ginsberg and R. Wickner, Fed. Proc. (Fed. Amer. Soc. Exp. Biol), 30, 1110 (1971).

[10] R. B. Wickner, B. Ginsberg, I. Berkower, and J. Hurwitz, J. Biol. Chem., 247, 489 (1972).

[11] R. B. Wickner, B. Ginsberg, and J. Hurwitz, <u>J. Biol. Chem.</u>, <u>247</u>, 498 (1972).

[12] R. Moses and C. C. Richardson, <u>Proc. Nat. Acad. Sci. USA</u>, <u>67</u>, 674 (1970).

[13] D. Brutlag, <u>Fed. Proc</u>. (Fed. Amer. Soc. Exp. Biol.), <u>30</u>, 1109 (1971).

[14] M. L. Gefter, I. J. Molineux, T. Kornberg, and H. G. Khorana, <u>J. Biol. Chem.</u>, <u>247</u>, 3321 (1972).

[15] J. L. Campbell, L. Soll, and C. C. Richardson, <u>Proc. Nat. Acad. Sci. USA</u>, <u>69</u>, 2090 (1972).

[16] Y. Hirota, M. Gefter, and L. Mindich, <u>Proc. Nat. Acad. Sci. USA,</u> <u>69</u>, 3238 (1972).

Chapter 7

DISSOCIATION OF PROTEIN AND RIBONUCLEIC ACID

SYNTHESIS IN VIVO

Herbert L. Ennis

Roche Institute of Molecular Biology
Nutley, New Jersey

I.   INTRODUCTION

In the steady-state growth of bacterial cells, the rates
of synthesis of RNA and of protein are normally maintained in
constant proportion by control mechanisms that are poorly
understood [1, 2]. These control mechanisms can be altered by
various methods that selectively inhibit protein or RNA synthesis.

The inhibition of protein synthesis, without a concomitant inhibition of RNA synthesis, results in an imbalance in the RNA
and protein content of the cell. Various methods are available
to dissociate RNA from protein synthesis. Among these, there
are treatment of cells with antibiotics or other compounds [3, 4],
amino-acid starvation of relaxed-control strains of Escherichia
coli [5], or K^+ accumulation [6]. The distortion of the normal
balance of cell constituents by these methods provides a suitable way of studying the interrelationships between the synthesis
of these macromolecules and the mechanisms involved in the regulation of the RNA and protein content of cells. In addition, the
formation of RNA without concomitant protein synthesis
enables one to study the production of all the types of RNA,
uncomplicated by simultaneous protein synthesis, and in this way
to determine the role of protein synthesis in regulating RNA
formation.

II. DESCRIPTION OF METHODS USED

A. Inhibition of Protein Synthesis by Antibiotics and Other Compounds

A large number of antibiotics have been isolated that
selectively inhibit protein synthesis in sensitive cells [7].
In general, cells incubated in the presence of any of these
antibiotics continue to synthesize RNA in the absence of concomitant protein synthesis. In the past, the most commonly
used antibiotics have been chloramphenicol and puromycin.

However, as the mode and site of action of other antibiotics become more fully understood, they will also undoubtedly be utilized, especially since many of these antibiotics inhibit protein synthesis at different steps in the biosynthetic pathway.

This method is simple and has the advantage that it can be used to inhibit protein synthesis in any sensitive organism. The antibiotic is added to cells growing exponentially in any of a large variety of media. If a sufficiently high concentration of drug is used, inhibition of protein synthesis is very rapid, being complete within seconds after addition of the compound. It is important to titrate the activity of the antibiotic against the organism being used, since some strains of bacteria are more resistant to drugs than others. In order to obviate any possible problems concerned with interpretation of the results obtained, it is important to use concentrations of drugs that rapidly and completely inhibit protein synthesis [8].

One drawback of using antibiotics is that sensitive strains of bacteria are required, and these are not always available. However, it is possible to isolate antibiotic-sensitive strains [9-11]. Another disadvantage is that if one wishes to reinitiate protein synthesis after antibiotic treatment, the cells must first be washed free of the drug and then inoculated into fresh medium lacking antibiotic. This takes time and, combined with any other toxic side-reactions induced by the antibiotic, may cause a lag in subsequent growth.

Another compound whose effect resembles that of antibiotics is 8-azaguanine, a guanine analogue. This compound also selectively inhibits protein synthesis in sensitive cells [12,13]. Although the drug is incorporated into RNA during growth of the cells in the presence of the compound, RNA synthesis is not inhibited. The RNA made in the presence of azaguanine, however, may be abnormal. The inhibition of protein synthesis by azaguanine is readily reversed by the addition of guanosine to the inhibited cultures. Consequently, protein synthesis can be easily restarted in inhibited cultures.

Cobalt added at suitable concentrations to growing cultures of E. coli has also been shown to inhibit protein synthesis selectively [14, 15].

B. Amino Acid Starvation of Relaxed and Stringent Bacterial Strains

Protein synthesis can be inhibited by depriving amino acid auxotrophs of E. coli of one or more of the required amino acids. Some strains of E. coli stop synthesizing stable RNA during starvation of a required amino acid, and are called stringent (rel$^+$). Certain mutant strains, however, continue the synthesis of stable RNA during such starvation and are called relaxed (rel$^-$) (see Ref. [16] for an extensive review of the literature). An example of these responses to amino acid starvation is given in Fig. 1. The results are given only for leucine starvation but, in general, are the same for starvation of other amino acids.

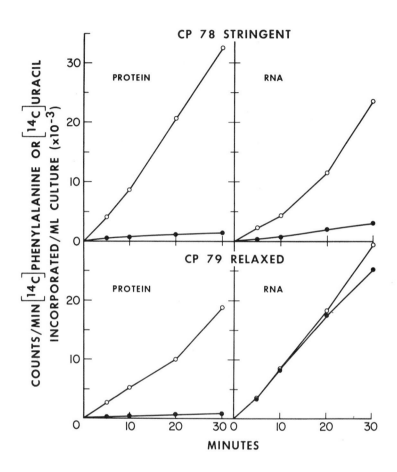

Fig. 1. The effect of amino acid starvation on [14C]uracil and [14C]phenylalanine incorporation into isogenic rel+ and rel− strains of E. coli. Cells growing exponentially at 37°C in mineral salts–glucose medium supplemented with all amino acids (100 μg/ml each) required by the bacteria plus thiamine (10 μg/ml) were centrifuged at 2000 x g for 5 min at room temperature. The

pellet was suspended in the mineral salts basal medium (lacking glucose and amino acids) and centrifuged again. The pellet was washed once more and the cells were suspended at 3.5×10^8 cells/ ml in 10 ml of the appropriate medium. Then [^{14}C] phenylalanine (0.1 μCi/ml or [^{14}C] uracil (0.1 μCi/ml) was added. The cultures were shaken vigorously at 37°C; 1-ml aliquots were taken at the indicated intervals and mixed with 1.5 ml of cold 10% trichloroacetic acid and kept in an ice bath for 30 min. To determine incorporation into nucleic acid, the samples containing [^{14}C] uracil were filtered through Millipore filters (0.45-μm pore size) and washed with four 5-ml portions of cold 5% trichloroacetic acid. The filters were dried and placed into scintillation fluid and counted with a Beckman LS-100 scintillation spectrometer. The samples containing [^{14}C] phenylalanine were heated at 100°C for 10 min to hydrolyze aminoacyl-tRNA, cooled, and then filtered and counted as above.

(A)   CP78 - stringent (<u>rel</u>$^+$). Left panel, [^{14}C] phenyla-line incorporation; Right panel, [^{14}C] uracil incorporation.

(0)   Complete medium: mineral - salts - glucose containing threonine, leucine, histidine, arginine, (100 μg/ ml, each), phenylalanine (10 μg/ml), uracil (10 μg/ml), and thiamine (10 μg/ml).

(●)   Starvation medium: as above except no leucine added.

(B)   CP 79 - relaxed (<u>rel</u>$^-$).

(0)   as (A)

(●)   as (A)

It is important to note that during amino acid starvation of either relazed or stringent strains, protein synthesis is never completely inhibited. Residual protein synthesis continues at about 5% of the exponential rate [17]. Consequently, when this technique is used to dissociate protein from nucleic acid synthesis, it is important to realize that some protein synthesis is taking place and to take this into consideration when the data are interpreted.

Although isogenic rel^+ and rel^- strains can readily be isolated [18], they have so far been obtained only in Gram-negative organisms. Consequently, this method cannot be used, at present, in certain species of bacteria. The method has the obvious advantage, contrary to the situation in which antibiotics are used, of restoring protein synthesis merely by adding back the amino acid to the starved cells.

Amino acid auxotrophs can be starved in a number of ways:

(i) Exponentially growing cells are washed 2 or 3 times in buffer (usually the basal salt medium of the complete growth medium), and then resuspended in warmed growth medium lacking the required amino acid. Incubation is then continued. RNA and protein synthesis are determined either by chemical methods [19, 20] or by the incorporation of radioactive precursors. (See the legend to Fig. 1 and [21] for a rapid and accurate method.)

The necessity for washing the cells free of amino acid before starvation can be eliminated by allowing the cells to deplete the growth medium of the required amino acid. In this case, the cells are grown in a medium containing a limiting amount of required amino acid. Growth, consequently, will stop due to exhaustion of the amino acid, and the cells can then be starved for the length of time desired.

(ii) Another useful method involves interference with the synthesis of amino acids by the use of analogues [22]. This method has the advantage that amino acid auxotrophs are not needed to do the experiment, and the inhibition can be readily reversed by adding large quantities of the natural amino acid. The analogues are merely added to growing cultures in sufficient quantity to inhibit protein synthesis. The following are some of the analogues that have been used: 5-methyltryptophan or β-indolacrylic acid [23] (tryptophan analogues), β-2-thienyla-lanine (aromatic amino acids), 2-thiazolealanine (histidine), and α-ketobutyrate (branched-chain amino acids).

Valine interferes with isoleucine synthesis in K12 strains of E. coli. Consequently, these strains can be starved for isoleucine by the addition of high concentrations of valine to growing cultures. Inhibition of protein synthesis is very rapid. Cysteine, which probably interferes with threonine synthesis, may also be used [24].

(iii) The drug trimethoprim inhibits the reduction of
dihydrofolate to tetrahydrofolate catalyzed by the enzyme dihydrofolate reductase. This reaction is a step in the synthesis of
N^{10}-formyltetrahydrofolate [25]. The latter is the source of
the formyl group in N-formylmethionyl-tRNA, the aminoacyl-tRNA
required for initiation of protein synthesis in E. coli. The
addition of trimethoprim to stringent strains of E. coli results
in the inhibition of both protein and RNA synthesis. However,
in relaxed strains although protein synthesis is inhibited RNA
formation continues, similar to what is observed during amino
acid starvation of these strains [26].

(iv) Another technique employed to dissociate protein from
nucleic acid synthesis in relaxed strains is amino acid starvation resulting from a transfer of a culture from rich medium containing a mixture of all the amino acids to a minimal medium
containing none [23, 24]. Repression of the enzymes responsible
for amino acid biosynthesis during growth in the rich medium
results in a transient amino acid starvation during this "shift-
down" to growth in mineral medium. Since the starvation is of
short duration, this technique has limited use.

(v) A number of conditional lethal mutants of E. coli
usable to synthesize protein at elevated temperatures have been
isolated. These mutants containing the rel^+ locus do not synthesize RNA at the nonpermissive temperature, whereas rel^- strains
do [27-29]. The method used is very simple. Cells are grown

exponentially at the permissive temperature (usually 30°C).
Then the culture is placed in a water bath at the nonpermissive
temperature (usually 40-42°C) and incubation is continued. It
was noted that protein synthesis in these strains stopped dram-
atically upon shift to the elevated temperature. RNA synthesis
was inhibited in the $\underline{rel}^+$ strain, but continued in the $\underline{rel}^-$
bacteria.

C. Potassium Depletion of Mutants Defective in Potassium Accumulation

Microbial cells require K^+ for growth and protein synthe-
sis. Mutants of $\underline{E. \; coli}$ B [6] and K12 [30], and $\underline{Bacillus \; sub-}$
$\underline{tilis}$ [31, 32] that have lost the normal capacity to transport
and concentrate K^+ from the growth medium cannot synthesize
protein when incubated in a medium lacking K^+. The internal K^+
concentration can be regulated by the use of these mutants and,
therefore, the rate of protein synthesis can be controlled by
variation of the external concentration of this cation. In
contrast to protein synthesis, RNA synthesis continues in K^+-
depleted cells.

The three strains can be depleted of their intracellular
K^+ by identical means. The following method has been used
routinely [32-34]:

Medium A (K^+-containing medium) for $\underline{E. \; coli}$ strains [35].

    7 g K_2HPO_4

    3 g KH_2PO_4

1 g $(NH_4)_2SO_4$

0.5 g Na_3citrate

48 mg $MgSO_4$

Distilled water to 1 liter

Add 0.25% glucose, final concentration (after autoclaving)

Medium A (K^+-containing medium) for B. subtilis strains.

Add 0.5% glucose, final concentration (after autoclaving)

0.1% Difco casamino acids

100 µg/ml tryptophan

25 µg/ml thymine

Medium Sodium A: As Medium A, with Na^+ phosphates substituted mole for mole for the K^+ phosphates. The media should be compounded of the best grade chemicals available. Contamination of chemicals with K^+ can result in growth of the mutants in Medium Sodium A. Uracil and adenosine (20 µg/ml of each) are usually added to Medium Sodium A, since we have found that these compounds greatly stimulate the RNA synthesis that occurs during K^+ depletion [36].

The strains can revert to the wild type if they are not grown properly. For best results we pick single colonies grown on Medium A agar plants and grow these colonies up overnight with vigorous shaking at 37°C, in about 25 ml of Medium A. These cultures may be kept in the refrigerator for as long as two weeks (in the case of B. subtilis, at room temperature for several days) and used as starter cultures. It is important to grow

the stocks on media containing glucose. In our limited experience, we have found that the strain reverts readily, for some
unexplained reason, when grown on glycerol for many generations.
The stock should be tested for growth in Medium Sodium A. Inoculate about 5×10^8 cells/ml (about 100 Klett units, number 42
filter) in Medium Sodium A and measure either Klett reading or
absorbance at 30-min intervals, or measure incorporation of $[^{14}C]$-
leucine into protein. At most, a 20-30% increase in Klett
reading is observed after 1-2 hr. No incorporation of $[^{14}C]$leucine
is seen. Compare this to the same bacterial strain inoculated
into Medium A. If growth occurs in Medium Sodium A, then either
the culture has reverted or the medium being used contains
enough K^+ to support growth. Care should be taken not to intro-
duce K^+ into the medium when other substances are used. For
example, if phage is added, make sure the stock is suspended in
Na^+ buffer.

To prepare cells for K^+ depletion, inoculate a culture
into any medium you desire to use that contains K^+ from the
stock previously grown on Medium A containing glucose. Let the
culture grow for three generations and harvest the cells during
the logarithmic growth phase. Centrifuge once at room temperature to sediment the cells. Decant the supernatant. Resuspend
the pellet in Medium Sodium A (without glucose) and centrifuge
again. Repeat this once more. Finally, resuspend the cells in
a small volume of Medium Sodium A (or any Na^+-containing medium

you may want to use) and use this as the inoculum into your K^+-free medium. One may also rapidly deplete the cells of K^+ by filtering and washing on 47-mm diam Millipore filters. Ten ml of a culture containing 5×10^8 cells/ml can rapidly be washed in this way. Pour the culture onto a moist filter (0.6-µm pore size) and apply suction. Wash two or three times with 10-ml portions of buffer (each).Then place the filter with the attached cells into a flask containing the medium to be used in that particular experiment. The cells will detach from the filter after a few seconds of rapid shaking. The filter may now be removed from the flask. This method gives almost quantitative recovery of cells. The entire procedure can be accomplished within 1 min. Its limitation is that only small quantities of cells can be manipulated. However, this method is excellent for studies involving incorporation of radioactive compounds since small numbers of cells can be used.

The kinetics of incorporation of $[^{14}C]$leucine into protein in cells growing in a medium containing various concentrations of K^+ is shown in Fig. 2. The rate of incorporation is markedly dependent upon the concentration of K^+ in the growth medium.

Figure 3 gives the results in terms of the intracellular K^+ concentration, and also shows that RNA synthesis occurs in the absence of K^+. Even when no protein synthesis is observed, RNA synthesis proceeds at approximately 50% of the exponential rate. The data presented in Fig. 3 are an average of 5 separate

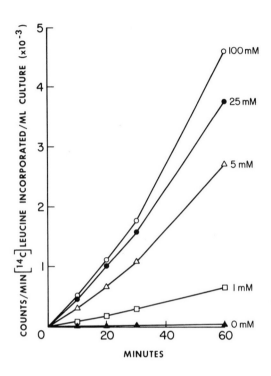

Fig. 2. The effect of potassium on [^{14}C]leucine incorporation in E. coli B207. Cells growing exponentially at 37°C in
medium A containing a complete mixture of amino acids, uracil,
and adenosine (20 µg/ml, each) were harvested, washed, and inoculated at 5 x 10^8 cells/ml into identical media containing the
indicated concentrations of K$^+$ (as described in the text). [^{14}C]-
Leucine (0.025 µCi/ml) was added to each culture, and incorporation of the compound into protein was determined as described
in the legend to Fig. 1. The data are presented for the K$^+$ concentration in the growth medium and cannot directly be compared

experiments. As can be seen from the standard error of the mean (brackets), the results are highly reproducible.

The usefulness of K^+ depletion as a method to dissociate protein from nucleic acid synthesis should be emphasized. K^+ depletion is a simple, rapid, and precise method to dissociate RNA from protein synthesis. The unique feature of this method is that one can maintain a very fine control over protein synthesis [32, 37] by the simple manipulation of the K^+ concentration in the growth medium. The ability to easily manipulate the rates of protein synthesis could provide a useful system for the future study of cellular control mechanisms.

Other methods available to inhibit protein synthesis are not completely adequate. During amino acid starvation of the relaxed or stringent strain, appreciable protein synthesis occurs [17]. During K^+ depletion, protein synthesis is inhibited by more than 99%. Cells treated with antibiotics to inhibit protein synthesis may not recover immediately from growth inhibition due to the drug. There are toxic side-reactions related to antibiotic action, and cells may be killed during the treatment. Furthermore, in order to restart protein synthesis after

---

Fig. 2 - continued

to the data presented in Fig. 3 without conversion to intracellular K^+ concentrations. For extracellular K^+ 100 mM, 25 mM, 5 mM, 1 mM, and 0 mM, the intracellular K^+ concentrations are approximately 190, 178, 125, 35, and less than 5, respectively.

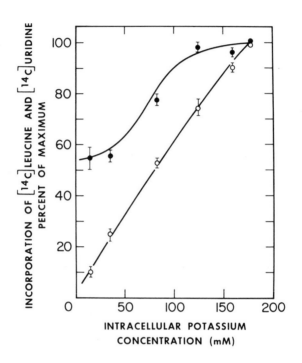

Fig. 3.   Incorporation of [^{14}C]leucine into protein (0-0)
and [^{14}C]uridine into RNA (0-0) at various concentrations of
intracellular K$^+$.   Temperature, 37°C.   After growth in minimal
medium (Medium A, 0.2% glucose, and histidine, leucine, and
methionine at 100 μg/ml, each) supplemented with uridine (40 μg/
ml), B207 cells were washed in Medium Sodium A and resuspended
in mixtures of A and Sodium A supplemented with 20 amino acids
at 40 μg/ml, guanine (25 μg/ml), adenine (40 μg/ml), uridine
(40 μg/ml), cytidine (40 μg/ml), and thymidine (20 μg/ml).
Intracellular K$^+$ concentrations were computed as previously

it has been stopped by the use of antibiotics, it is necessary
to wash the cells free of the drug. This takes time and is
likely to induce a lag in subsequent growth. Very high concentrations of antibiotic must be used to completely inhibit protein synthesis [8]. This disadvantage may be overcome by the
use of the K^+ accumulation-deficient mutant. K^+ depletion is
rapid and simple to perform. Cells may be washed rapidly on
large Millipore filters and then resuspended in the appropriate
medium lacking K^+. The whole operation can take less than 1 min.
One need only restore K^+ to start protein synthesis. Bacteria
can be deprived of K^+ for long periods of time and yet retain
excellent metabolic activity when K^+ is restored. This method
of inhibition of protein synthesis would seem to be more physio-
logical than the use of antibiotics.

III. CONCLUDING REMARKS

It is important to realize that all the methods that can
be used to inhibit protein synthesis are probably not identical.
Not only is protein synthesis inhibited at different steps by
the various methods, but the physiology of the inhibited cell

---

Fig. 3 - continued
described [37]. Each point shown represents rate of synthesis
after transfer to new medium. Bracket shows $1 \pm$ S.E. of mean
(n = 5). All values expressed as percent of the rate on the
medium with the highest K^+ concentration. (Figure from [37],
reproduced with permission from the publisher).

is not equivalent. For example, it is becoming increasingly
evident that not all classes of RNA are made during inhibition
of protein synthesis by the methods described. In the presence
of chloramphenicol, no β-galactosidase messenger RNA can be
induced [38], whereas during amino acid starvation of $\underline{rel}^-$ and
$\underline{rel}^+$ cells, this species of RNA is made [39]. The strain used
is also important, because the effect of chloramphenicol on
the synthesis of β-galactosidase messenger can be suppressed
by a SuA mutant, i.e., β-galactosidase messenger is made during
antibiotic treatment of a SuA strain [38]. Furthermore, during
chloramphenicol inhibition and starvation of $\underline{rel}^-$ or $\underline{rel}^+$ strains,
some residual protein synthesis occurs [8, 17].

However, with these limitations in mind, these methods
have proved valuable in the past, and undoubtedly will be useful in the future for studies of the interrelationship between
protein and nucleic acid synthesis in bacterial cells.

REFERENCES

[1] M. Schaechter, O. Maaløe, and N. O. Kjeldgaard, J. Gen.
Microbiol., 19, 592 (1959).

[2] F. C. Neidhardt and B. Magasanik, Biochim. Biophys. Acta,
42, 99 (1960).

[3] E. F. Gale and J. B. Folkes, Biochem. J., 53, 493 (1953).

[4] B. H. Sells, Biochim. Biophys. Acta, 80, 230 (1964).

[5] E. Borek, A. Ryan, and J. Rockenbach, J. Bacteriol., 69,
460 (1955).

[6] H. L. Ennis and M. Lubin, Biochim. Biophys. Acta, 50, 399 (1961).

[7] B. Weisblum and J. Davies, Bacteriol. Rev., 32, 493 (1968).

[8] A. I. Aaronson and S. Spiegelman, Biochim. Biophys. Acta, 53, 70 (1961).

[9] D. Apirion, J. Mol. Biol., 30, 255 (1967).

[10] M. Sekiguchi and S. Iida, Proc. Nat. Acad. Sci. USA, 58, 2315 (1967).

[11] H. L. Ennis, J. Bacteriol. In press.

[12] H. Chantreene and S. Devreux, Exp. Cell Res. Suppl., 6, 152 (1958).

[13] H. Chantreene and S. Devreux, Biochim. Biophys. Acta, 41, 239 (1960).

[14] H. B. Levy, E. T. Skutch, and A. L. Schade, Arch. Biochem., 24, 199 (1949).

[15] M. R. Blundell and D. G. Wild, Biochem. J., 115, 213 (1969).

[16] G. Edlin and P. Broda, Bacteriol. Rev., 32, 206 (1968).

[17] D. Goodman, J. Mol. Biol., 51, 491 (1970).

[18] N. Fiil and J. D. Friesen, J. Bacteriol., 95, 729 (1968).

[19] W. C. Schneider, J. Biol. Chem., 161, 293 (1945).

[20] O. H. Lowry, N. J. Rosebrough, A. L. Farr, and R. Randall, J. Biol. Chem., 193, 265 (1951).

[21] D. B. Roodyn and H. G. Mandel, Biochim. Biophys. Acta, 41, 80 (1960).

[22] M. H. Richmond, Bacteriol. Rev., 26, 398 (1962).

[23] D. E. Morse, R. Mosteller, R. F. Baker, and C. Yanofsky, Nature, 223, 40 (1969).

[24] F. C. Neidhardt, Biochim. Biophys. Acta, 68, 100 (1963).

[25] J. J. Burchall and G. H. Kitchings, Mol. Pharmacol., 1, 126 (1966).

[26] A. Y. Shih, J. Eisenstadt, and P. Lengyel, Proc. Nat. Acad. Sci. USA, 56, 1599 (1966).

[27] L. Eidlic and F. C. Neidhardt, J. Bacteriol., 89, 706 (1954).

[28] A. Bock, L. E. Faiman, and F. C. Neidhardt, J. Bacteriol., 92, 1076 (1966).

[29] G. P. Tocchini-Valentini and E. Mattoccia, Proc. Nat. Acad. Sci. USA, 61, 146 (1968).

[30] H. L. Ennis, Unpublished observations using strain 2K118 isolated by W. Epstein, University of Chicago.

[31] M. Lubin, Fed. Proc., 23, 994 (1964).

[32] D. B. Willis and H. L. Ennis, J. Bacteriol., 96, 2035 (1968).

[33] M. Lubin and D. Kessel, Biochim. Biophys. Res. Commun., 2, 249 (1960).

[34] P. S. Cohen and H. L. Ennis, Virology, 27, 282 (1965).

[35] B. D. Davis and E. S. Mingioli, J. Bacteriol., 60, 17 (1950).

[36] H. L. Ennis and M. Lubin, Biochim. Biophys. Acta, 95, 605 (1965).

[37] M. Lubin and H. L. Ennis, Biochim. Biophys. Acta, 80, 614 (1964).

[38] H. E. Varmus, R. L. Perlman, and I. Pastan, Nature New Biology, 230, 41 (1971).

[39] M. Artman and H. L. Ennis, Unpublished observations.

Chapter 8

CELL-FREE SYNTHESIS OF VIRAL PROTEINS IN THE

KREBS II ASCITES TUMOR SYSTEM

I. Boime[*] and H. Aviv

Laboratory of Molecular Genetics
National Institute of Child Health and Human Development
National Institutes of Health
Bethesda, Maryland

[*]Present address: Departments of Obstetrics, Gynecology, and
Pharmacology, Washington University School of Medicine,
St. Louis, Missouri   63110

I. INTRODUCTION

Our understanding of the processes involved in bacterial protein biosynthesis has been considerably advanced by the development of very active cell-free systems in which specific proteins are synthesized in response to specific, naturally occurring, mRNAs. With these systems it is possible to detail three reaction sequences involving the initiation, elongation, and termination of specific bacterial proteins. Until recently, however, progress in elucidating these translational reactions and understanding their regulation in animal cells has been hampered by the lack of an efficient cell-free system that responds accurately to the addition of natural mRNAs. Clearly a mammalian cell-free system that translates added viral and cellular mRNAs into authentic identifiable products has obvious advantages for the study of the processes of translation and gene expression in higher organisms. Development of the system we describe was poineered by T. S. Work and his associates [1, 2], and more recently advanced by Mathews and Korner [3].

RNA derived from the mammalian encephalomyocarditis (EMC) virus directs protein synthesis in extracts from Krebs II ascites

tumor cells propagated in mice [1-4]. The products of this
reaction correspond to authentic viral polypeptides [5, 6]. To
establish this system in our laboratory, we have adopted the
procedure of Martin et al. [7], of Kerr et al. [2], and of
Mathews and Korner [3] for preparation of ascites tumor cell
extracts. The system has been modified so that it translates
EMC RNA very efficiently _in vitro_ into authentic viral products
[8]. Here, we describe in detail how the system is prepared.
In addition, we discuss the advantages, as well as the possible
uses of the system in examining protein biosynthetic mechanisms
and its control in animal cells.

II. PRODUCTION OF ASCITES TUMOR CELLS

An ascites tumor is a tumor that results from active
multiplication of free neoplastic cells in peritoneal fluid of
the abdominal cavity, leading to very high concentrations of
tumor cells, and yielding a nearly pure culture [9]. "The ascites
tumor is basically a pathological condition of the peritoneum
resulting from the leakage of serous fluid from small veins and
and capillaries in the peritoneal subserosa after infiltration
with tumor cells (provoking anoxia and low-grade inflammation).
This leakage occurs wherever intraperitoneally injected cells
and fragments of malignant tissue find suitable anatomic conditions for abundant implantation into the peritoneum [10]."

There are many lines of tumor cells that will grow in the
ascitic form. The line we have used in our studies is the Krebs II

ascites tumor, which is derived from a solid mammary carcinoma [9], and seems to propogate only in mice. In addition to their vigorous growth characteristics, these cells have the advantage of growing in noninbred strains of mice. We maintain a stock of frozen cells as a standard preparation, and build up fresh experimental stock from these cells as needed.

Male mice (20-25 g) are used in all of our studies. The strain used is the N.I.H. General Purpose mouse, a breed derived from the Swisstar strain. The animals are injected intraperitoneally with 0.1 - 0.2 ml of freshly harvested, unwashed ascites cells. The cells used for injection are from mice that developed large yields of blood- and clump-free ascites fluid.

The cells are harvested 7-8 days after injection. The mice are killed by cervical dislocation and the abdomen is washed with 70% ethanol; the entire animal is then rinsed under running tap water. This latter step is added to aid in the removal of urine and fecal material, as well as loose hair, that might be carried along with the ascites fluid during collection into the wash buffer. The skin and peritoneum are cut carefully so that the intestines are not perforated, and the cells are collected through two layers of cheesecloth into a vessel containing the appropriate cold-wash buffer. Each mouse yielded approximately 5- to 10-ml of fluid containing 1×10^8 cell/ml. If, as seen through the peritoneal membrane, the ascitic fluid

is bloody or contains large clumps, such fluid is not collected.
It is important that the cells be harvested not later than 8
days after injection. After this time, the ascitic fluid develops large clumps of fatty debris and nonviable ascites cells.
Such older, turbid fluids were significantly less effective for
inoculation.

Ascites tumor cells can be frozen for subsequent injection,
but the freezing procedure must be carefully controlled. Cells
washed with phosphate-buffered saline (Buffer A) should be frozen
in such a way that the rate of temperature reduction to about
-60°C is 1°C/min. When the temperature reaches -60°C the samples
are transferred to liquid nitrogen. There are special cryogenic
devices that can be used for this procedure, for example, the
Linde B. F. 4 Freezing System. In the absence of such a freezing apparatus, the cells can be placed on dry ice until they
are frozen and then stored in liquid nitrogen. However, the
take rate (animals developing tumor/animals injected) is reduced to about 60%, and the number of stained cells (cf. III A),
which reflect losses of cell viability, is increased from the
normal 5 to 10% to about 50% of the total cells. Whatever
freezing techniques are used, sufficient glycerol to give a
final concentration of 6% is added to the cells in order to increase the resistance of the cells to freezing. For a further
discussion of the above topics, see Martin et al. [7].

III. PREPARATION OF CELL-FREE EXTRACTS FROM ASCITES TUMOR CELLS

A. Preparation of an Extract

All solutions are prepared with sterile water and all glass-ware is autoclaved. Fresh cells are washed three times by cen-trifugation (60 x g, 4 min) with cold buffer containing 35 mM Tris·HCl (pH 7.5)-140 mM NaCl (buffer B). The cells are diluted with an equal volume of buffer B, and the cell concentration is measured by dilution of the cells 1:100 in 0.1% Trypan blue dissolved in buffer A. The cells are then counted in a Neubauer heamocytometer. Nonviable cells are stained blue and are not included in the count; they should constitute no more than 10% of the total cells observed in a low-power microscopic field. The cells are finally packed by centrifugation at 250 x g for 5 min, resuspended in two packed-cell volumes of hypotonic buffer containing 10 mM Tris·HCl (pH 7.5)-10 mM KCl-1.5 mM Mg $(CH_3COO)_2$ (buffer C), and homogenized with about twenty strokes of a tightly fitting Dounce homogenizer. The homogenate is then brought to a final concentration of 30 mM Tris·HCl (pH 7.5)-125 mM KCl-5 mM Mg $(CH_3 COO)_2$-7 mM 2-mercaptoethanol (buffer D) by the addi-tion of 1/9 volume of a 10-fold concentrated solution, and cen-trifuged at 30,000 x g for 10 min. The supernatant fluid is collected with a pipette; care should be taken not to collect the pellicle of lipid on the surface. The pellet is discarded. After adjustment of the extract to final concentrations of 1 mM ATP, 0.1 mM GTP, 0.6 mM CTP, 10 mM creatine phosphate, 0.16 mg/ml

of creatine kinase, and 40 µM (each) of 20 amino acids, it is incubated at 37°C for 45 to 60 min and centrifuged at 30,000 x g
for 10 min. The pellet is discarded. In order to remove amino
acids and bring the concentration of the extract to that of
buffer D, the extract is passed through a Sephadex G-25 column
previously equilibrated with this buffer. The volume of the
column used should be at least 10-fold greater than that of the
extract. Alternatively, the extract can be dialyzed against
this same buffer for 6 hr. At this stage, its A_{260}/ml value
should be 35 to 60. The extract was either aliquoted and frozen
in liquid N_2, or was used for the preparation or ribosomes (see
III, B).

The above preincubation step is necessary to obtain an
efficient mRNA-dependent system. Incubation lowers the background synthesis activity by allowing ribosomes to complete the
synthesis of nascent peptide chains, with their subsequent
release from endogenous mRNA. The mRNA released is presumably
degraded.

B. Preparation of Ribosomes and Ribosome-Free Extracts

To prepare ribosomes, the preincubated extract is layered
on a solution of Buffer D containing 1.0 M sucrose for centrifugation in a Spinco #30 rotor. The amount of the buffered sucrose
solution used was one-fourth of the volume of the centrifuge
tube, and centrifugation was at 30,000 rpm for 12 hr at 4°C.
After centrifugation, the upper portion of the supernatant fluid,

containing lipid, is removed. The remainder is saved for use
as a ribosome-free extract. This material is either dialyzed
or passed through a Sephadex G-25 column as described above.

The resulting colorless ribosomal pellets were resuspended
in small volumes of Buffer D (1-2 ml/tube) with the aid of a
hand homogenizer. Their final concentration is 100-200 A_{260}/ml.
The suspended mixture is then centrifuged at 10,000 x g to remove aggregated material. Both ribosomes and S-100 are frozen
in 0.1- to 0.2-ml aliquots and stored in liquid nitrogen.

C. Preparation of tRNA from Ascites Cells

All glassware and solutions used in preparing all RNA
fractions were sterilized. Ascites cell tRNA was prepared by
a modification of the procedure described by von Ehrenstein (11).

1. tRNA Extraction by Phenol. 100 g of washed cells,
fresh or frozen, are suspended in 300 ml of Buffer B. An equal
volume of redistilled phenol (saturated with buffer B) is added
to the suspension. Water-saturated redistilled phenol can be
stored in amber bottles at 4°C for several weeks. The extraction is carried out with the aid of a magnetic stirrer for 30
min at 5°C. The phenol mixture is centrifuged at 10,000 x g
for 20 min, and the phenol phase is re-extracted with water and
again centrifuged. The combined aqueous phases are re-extracted
with phenol; to the resulting aqueous phase is added 1/9 volume
of 20% potassium acetate (pH 5.5) and 2 volumes of ethanol.
The ethanolic solution is left overnight at -20°C. The RNA is

then pelleted by centrifugation at 10,000 x g for 10 min.

    2. Sodium Chloride Extraction. The pellet is extracted twice by vigorous stirring with 50 ml of 1 M NaCl for 1 hr at 5°C. The washed suspension is centrifuged at 10,000 x g for 10 min, and the supernatant is saved. The resulting pellets are re-extracted with 25 ml of 1 M NaCl and centrifuged; the supernatants are pooled. The extracted material is precipitated with 2 volumes of ethanol. The pellet obtained after this step contains mainly ribosomal RNA and some DNA.

    3. Isopropanol Precipitation. The ethanol-precipitated material is pelleted and dissolved in 30 ml of 0.3 M sodium acetate (pH 7.0); 0.54 volume of isopropanol is added dropwise to the RNA solution at 5°C. The isopropanol-treated solution is transferred to a water bath and incubated at 23°C for 10 min with continual stirring. The RNA suspension is then centrifuged at -20°C at 10,000 x g for 10 min. The pellet is dissolved in 10 ml of 0.3 M sodium acetate (pH 7.0). The preceding steps are repeated with 0.54 volume of isopropanol, the supernatants are combined, and the final volume is recorded. The pellet obtained after the second fractionation was discarded.

    Based on the final volume of the combined supernatant fluids, an additional 0.44 volume of isopropanol is added dropwise to the total RNA solution (total amount of isopropanol added was then 0.98 volume), and the resulting solution is centrifuged at 10,000 x g for 10 min. The final pellet obtained, which contains the tRNA, is dissolved in 3 ml of water containing 2% potassium acetate (pH 5.5). Two volumes of 95% ethanol are added to the dissolved tRNA, and the RNA is allowed to precipitate overnight at -20°C. The tRNA is pelleted and washed 2 times with 70% ethanol containing 2% potassium acetate (pH 5.5) by centrifugation at 10,000 x g for 10 min. After these washings, the pellet is dissolved in enough water to give a final tRNA concentration of about 100-150 A_{260}/ml. This preparation contains primarily 4S RNA. The isopropanol step is included to remove additional ribosomal and messenger RNA.

One problem frequently encountered in the preparation of tRNA is the large amount of material that accumulates at the phenol-water interphase. This leads to difficulties when collecting the aqueous phase, since a significant amount of interfacial material will be carried over. To reduce an excessively large interphase, 10% chloroform can be added to the phenol before the initial extraction procedure.

It is important to realize when preparing tRNA, as with mRNA, that there is always the problem of nuclease activities that might contaminate the preparation from unsterilized glassware, water, and fingers. Therefore, it is always a good practice to sterilize all glassware and reagents, and to wear rubber gloves when preparing RNA. In addition, when RNA is stored as an alcoholic precipitate at low pH, it is less susceptible to nuclease attack.

IV. PREPARATION OF EMC VIRUS AND EMC RNA

A. Growth of Virus

Ascites cells not only serve as a source for the components of the cell-free system, but also are a suitable host cell for the growth of encephalomyocarditis (EMC) virus. This virus, like that of polio, is a member of the picornavirus group. The EMC genome is a single-stranded RNA having a molecular weight of 2.5×10^6 [12]. The virus completely inhibits all host macromolecular synthesis within a few hours after infection [13]; thus, after infection, essentially all protein synthesis taking place in the host cell is viral-directed.

Virus was grown on Krebs II ascites tumor cells, essentially as described by Bellett and Burness [14] and Martin et al. [7]. Ascites cells are harvested and washed as described above, except that the wash buffer used to collect the cells is buffer A. After they are washed, the cells are resuspended in an equal volume of buffer A and their concentration is determined (see III A). The cell concentration at this point should be about 10^8/ml.

The cells are diluted to a final concentration of 10^7/ml with Earle's Balanced Salt Solution (buffer E) containing 1% heat-inactivated horse serum and 0.22% $NaHCO_3$. Buffer E can be purchased as a 10-times concentrated stock solution from various biological supply companies. Thus, 100 ml of incubation mixture required 10 ml of 10-times Buffer E plus 84 ml of sterile H_2O,

5 ml of 4.4% sodium bicarbonate (previously gassed with CO_2),
and 1 ml of heat-inactivated horse serum (inactivated for 30 min
at 56°C). With this reconstituted solution, the ascites cells
are diluted to 10^7 cells/ml, allowing for the volume of virus
to be added later.

The amount of virus to be added is computed on the basis
of the desired multiplicity of infection (viral particles added
per host cell). To calculate the volume of virus needed to
obtain a given multiplicity, the total number of cells in a
culture are divided by the titer of the virus (expressed as
hemagglutination units or plaque-forming units/ml, see IV, B)
and multiplied by the required multiplicity. For example, given
a 1000-ml culture of 10^7 cells/ml and a virus stock solution
having a titer of 5 x 10^8 PFU/ml, the volume of virus necessary
to give a multiplicity of 3 PFU/cell would be:

$$\frac{1000 \text{ ml} \times 10^7 \text{ cells/ml} \times 3}{5 \times 10^8 \text{ PFU/ml}} = 60 \text{ ml}$$

Therefore, the infected culture consists of 100 ml of 10^8 cells/
ml, 60 ml of virus, and 840 ml of the buffer E mixture. One
procedure that we have recently used is to preincubate the un-
diluted cells with the requisite amount of virus for 1 hr at
5°C (A.T.H. Burness, personal communication). Inclusion of this
step in the viral preparation appears to increase the viral
titer almost 2-fold.

The infected cell suspension is then dispensed in spinner bottles and stoppered. The culture fluid added must not exceed 10 to 15% of the total volume of the bottle. This is a significant point, since virus yields are greatly reduced when the volume of fluid exceeds these limits. A magnetic stirring bar is then placed at the bottom of the container to maintain the cells in suspension. Also added to the culture are 100 units/ml of penicillin and 100 µg/ml of streptomycin sulfate. The infected cultures are then incubated at 37°C on magnetic stirrers, which are set at low speed. An uninfected control should be included in these experiments.

Intracellular virus titer reaches a peak at 7 to 8 hr after infection, and the extracellular titer reaches a maximum about 12 hr after infection. The cells are harvested 18 hr after infection. When infection is complete, it is almost always observed that the suspension is very turbid and contains large cellular aggregates. In addition, more than 80% of the cells are stained, whereas in the uninfected control the number of stained cells is 10 to 20%. An often-encountered problem when preparing virus is the formation of cellular aggregates within 3 hr after infection. Usually the aggregates will resuspend about 2 hr later. If the clumps remain throughout the duration of the incubation, the titer will be greatly reduced. The reason for this clumping is not understood. The infected culture fluid is centrifuged at 10,000 x g for 15 min at 4°C, and

the supernatant is either frozen as a source of viral stock or
is processed further in order to obtain purified virus. The
pellet is also saved, since it contains a considerable amount
of virus (see below).

B. Determination of Viral Titer

The viral titer can be determined by a plaque assay or by
estimation of hemagglutinating activity of the virus. We use
the hemagglutination assay because it is much quicker and is as
reliable as the plaque assay.

EMC virus agglutinates sheep erythrocytes, and the viral
hemagglutinating activity is a physical property of the virus
[15]. Fresh sheep erythrocytes (contained in Alsever's solution)
are washed three times by centrifugation at 2,000 rpm in a
solution containing 1 part of buffer A containing 0.01% $CaCl_2$
and $MgCl_2 \cdot 6H_2O$ (PBS), 1 part 4.5% glucose, and 0.10 part of 1%
gelatin. After the final washing, the volume of packed cells
is recorded and the cells are made up to 0.1% in this medium.

The serology trays used for the assay are designated
"Microtiter" "V" (see Appendix) plates, and the cups have a
capacity of 125 µl. 50-µl aliquots of the above medium are
placed in all the cups, To the first cup is added 50 µl of a
virus lysate. Then with "Microdiluters" (see Appendix) the
virus is diluted, starting with well #1, to a dilution of 1/2048
or more. In addition to estimating the titer of the experimental lysate, diluent alone should be placed in a row of wells,
in addition to a dilution of a virus pool of known titer.

The hemagglutination trays are then incubated at $3°C$ for
at least 4 to 6 hr. At this time, it is possible to detect
hemagglutination activity, which is manifested by the presence
of a "red button" at the bottom of the well. The end point of
each sample is the highest dilution showing a trace of hemagglutination, and the titer is usually expressed as the reciprocal
of this dilution in terms of hemagglutination units/ml (H.A./ml).
For example, a 50-μl aliquot of a viral lysate hemagglutinates
upon dilution to nine cups or 1/512 dilution. This is equal to
512 H.A. units /50 μl or 1.024×10^4 H.A. units/ml. This approxi-
mates 1.024×10^9 PFU/ml. (1 H.A. unit is equivalent to about
10^5 PFU).

C. Purification of Virus

Burness [16] has described a useful method for purification of EMC that we have used to produce very active preparations of viral RNA. Since the availability of pure viral RNA
limits the rate at which the system can be used, we have altered this procedure to simplify it and increase its yield.

The clarified viral supernatant is collected and immersed
in an ice bath. Sodium acetate (10 ml/liter of a 20% solution)
is added, followed by the dropwise addition of 2 N acetic acid.
The supernatant is titrated to a pH of 5.0, and stirred for an
additional 15 min. The suspension is centrifuged at 10,000 x g
for 10 min at $4°C$, and the well-drained pellets are collected.
The precipitate, containing the EMC virus and various cellular

components, is resuspended in buffer F (20 ml/liter of original
culture fluid, see Appendix) with the aid of a Dounce homogenizer with a loose-fitting pestle. The suspension at this stage
can be stored for several months at -20°C without loss of virus
titer.

The concentrated virus is then diluted 1:6 in sterile
distilled water, and 30 ml of this solution is layered over
7 ml of a CsCl solution (density 1.39) and centrifuged at
25,000 rpm for 6 hr at 10°C in a Spinco SW-27 rotor. The virus
forms a band just beneath the CsCl-lysate interphase, and is
collected by puncture of the tube bottom. The band can usually
be collected in one fraction, and the amount of virus is determined by hemagglutination titration. There is always a
large amount of protein that accumulates at the interphase that
could trap the virus or impede the sedimentation of the virus
into the CsCl solution. Therefore, this fraction also should
be titered, diluted further, and centrifuged again if a significant titer is present in this interfacial fraction. This
problem can be avoided if the time of the initial centrifugation
is increased.

The density of the virus fraction collected from the first
CsCl centrifiguation is adjusted to 1.33 g/ml, and centrifuged
at 50,000 rpm at 10C° for 20 to 24 hr in an SW-56 rotor. A sharp
viral band is formed in the CsCl gradient at a density of about
1.33 g/ml. This was collected in drops from the bottom of the

tube, and dialyzed against a solution containing 10 mM Tris·HCl (pH 7.5)-100 mM NaCl. If the virus is to be stored, it is made 1 mM in dithiothreitol and stored in liquid nitrogen.

An additional 30 to 50% of the virus could be recovered from the pellet obtained after centrifugation of the crude infected lysate. To purify the virus in this fraction, a small volume of PP8 buffer containing 1% NP-40 or BRIJ-38 (nonionic detergents) is added, and the pellet is rehomogenized with a Dounce homogenizer (tight pestle). The suspension is centrifuged at 8,000 x g for 15 min, and the supernatant is directly layered onto the CsCl solution and purified as described.

Purification of EMC with CsCl has advantages over previously published methods. It avoids an organic-phase extraction, which we found resulted in an 80% loss of material, and it permits an overall recovery of purified virus of 50 to 75%, as measured by hemagglutination titration. The new procedure is relatively fast and requires only reagents that are both stable and commercially available. Another rapid procedure to purify EMC, which involves polyethylene glycol precipitation, is beyond the scope of this chapter [28].

D. Purification of EMC RNA

It must be stressed again that for all operations in the purification of the RNA, only sterile glassware and solutions should be used. The RNA is prepared from purified dialyzed EMC virus by the method of Kerr et al. [2]. The concentration of

the dialyzed virus to be extracted should be at least 15 A_{260}/ml.
Usually a slight precipitate formed after dialysis. This material was not removed. EDTA is added to the viral mixture to a
final concentration of 1 mM, and the mixture is then equilibrated
at 43°C for 2 min. Deoxycholate is then added to give a final
concentration of 0.5%; immediately, 0.5 volume of phenol (prewarmed to 43°C) saturated with 50 mM Tris·HCl, pH 7.5, and 1 mM
EDTA is added. The suspension is then mixed with the aid of a
Pasteur pipette and centrifuged at 9000 rpm for 10 min. The
upper aqueous phase is carefully removed. The lower phenol layer
is washed with 0.5 volume of water, and the mixture is centrifuged
again at 9,000 rpm for 10 min and the aqueous phases are pooled.
One-ninth volume of 20% potassium acetate, pH 5.4, and 2 volumes
of ethanol (chilled to -20°C) are added to the aqueous material,
and the mixture is incubated overnight at -20°C.

The RNA is centrifuged at 9,000 rpm for 10 min and the
pellet is washed three times with 2 ml of a solution containing
2 parts of ethanol and 1 part of 0.15 M NaCl; with each wash, the
RNA is pelleted at 9,000 rpm for 10 min at 4°C. After the final
wash, the pellet is well drained and dissolved in enough water
to give an RNA concentration of 10 to 20 A_{260}/ml. The RNA is
aliquoted in small volumes and stored at -20°C.

V.   CONDITIONS FOR AMINO ACID INCORPORATION

Each 50-μl reaction mixture contains 30 mM Tris·HCl (pH
7.5), 5 mM (for EMC RNA) magnesium acetate, 120 mM KCl, 7 mM

2-mercaptoethanol, 1 mM ATP, 0.1 mM GTP, 0.6 mM CTP, 10 mM

creatine phosphate, 0.16 mg/ml creatine kinase, 40 μM (each of)

19 nonradioactive amino acids, 5 μM radioactive amino acid of

indicated high specific activity ([^{3}H]alanine, [^{14}C]phenylalanine,

or [^{35}S]methionine), 0.1 to 0.2 A$_{260}$ units of tRNA prepared from

ascites tumor cells, and 0.02 to 0.1 A$_{260}$ units of EMC RNA. In

addition, preincubated extract, or ribosomes and ribosome-free

extract prepared from ascites cells, are added. The amounts of

these used are variable, and concentration curves should be pre-

pared to determine how much of those components should be used.

The amount of extract usually added is 150-200 μg of protein.

For ribosome-free extract and ribosomes, 100 to 200 μg of protein

and 0.05 to 0.2 A$_{260}$, respectively, added to the above incorpor-

ation mixture, are satisfactory.

Incubation is at 37°C for 60 to 150 min. Reactions are

stopped by the addition of 0.2 ml of 0.1 M KOH. Incubation is

continued for 20 min, and 1 ml of ice-cold 10% Cl$_3$CCOOH is

added. The mixture is cooled at 0°C for 5 min, and the precipi-

tate is collected over a 0.45-μm Millipore filter, washed 3

times with 3 ml (each) of 5% Cl$_3$CCOOH, dried, and counted in a

liquid scintillation counter. All determinations are done in

duplicate.

The alkaline hydrolysis step is included to hydrolyze the

amino acid and the nascent peptide chains attached to tRNA.

Since the reaction mixtures contain labeled free amino acid,

there would be substantial amounts of labeled aminoacyl-tRNA
that would be precipitated by Cl_3CCOOH along with the proteins.

When different labeled amino acids or different batches
of the same labeled amino acids are used, the system should be
optimized for the respective isotope. A cocktail can be maintained for several weeks containing Tris·HCl, magnesium acetate,
KCl, ATP, GTP, CTP, creatine phosphate, and unlabeled amino
acids. Creatine kinase, 2-mercaptoethanol, labeled amino acid,
tRNA, and EMC RNA are freshly mixed as needed.

We have found, as others have, that the Mg^{2+} optimum for
EMC translation is 4 to 5 mM [3]. This is a sharp dependence,
since at 2.5 mM and 8 mM there is no apparent EMC translation.
We have also found that both the Mg and the KCl optimum varies
for different preparations of extracts. Therefore, each
preparation should be optimized with respect to Mg and KCl
concentrations.

A. Stimulation by EMC RNA

At saturating concentrations of tRNA, the system is
extremely efficient in its ability to translate viral RNA.
For example, when the incorporation of [3H]alanine into Cl_3CCOOH-
precipitable protein was determined, it was found that 0.02 A_{260}
(1.0 µg) of EMC RNA directed a greater than 20-fold response in
[3H]alanine incorporation. At nearly saturating concentrations
of EMC RNA (0.1 A_{260}), the effect is more than 40 times the
background value. In terms of counts, when [3H]alanine with a

specific activity of 40 Ci/mmole is used, ^{3}H incorporation
represents 200,000 counts in the presence of EMC RNA, over a
minus-message background of 4,000 - 5,000 counts. This message-
dependent synthesis is equivalent to the incorporation of about
12 pmoles of [^{3}H]alanine.

By the use of two-dimensional fingerprint techniques, it
has been shown that portions of the EMC genome corresponding to
authentic structural proteins of the virus are translated by
the ascites cell-free system [5, 6]. In addition, about one-
third of the information encoded by the EMC genome is translated in vitro as a single polypeptide chain. Thus, the high
efficiency of message-dependent incorporation of amino acid
reflects in the product analysis that a long stretch of the
EMC genome is translated in vitro in appropriate phase.

B.   tRNA Requirement

We have found that the dependence of viral RNA-directed
synthesis upon the addition of tRNA varies among different extracts. Some extracts are almost entirely dependent upon tRNA
addition, while others show only 3- to 4-fold stimulation upon
the addition of tRNA. What accounts for this variation is not
clear. In addition, there seems to be a limited range of species
from which tRNAs can be derived to support protein synthesis in
the ascites tumor cell system. In addition to tRNA derived from
ascites tumor cells, tRNA prepared from rat liver, beef brain,
and duck liver support viral RNA-directed amino acid incorporation.

In contrast, neither tRNA derived from yeast nor tRNA derived from E. coli permits significant amino acid incorporation. These findings have a practical significance, since some species of tRNA that can be used in the cell-free system are commercially available, e.g., rat liver tRNA.

C. Time of Incubation with EMC RNA

As mentioned previously, the products synthesized in the cell-free system programmed with EMC RNA correspond to authentic viral capsid proteins. However, the higher molecular weight proteins appear clearly only after an incubation period of 90 min. Although the amount of amino acid incorporated into Cl_3CCOOH-precipitable protein, with respect to time, is linear for about 60 min, there still is significant incorporation from 60 to 150 min. Therefore, at least in terms of EMC RNA-directed synthesis, the appearance of discrete message-dependent products is dependent on the period of incubation.

VI. CONCLUSIONS

Two advantages of the viral RNA-ascites tumor cell system are that the virus can be grown to a high titer in dense cell suspensions and that the cells need not be maintained by delicate tissue culture procedures. In addition, cell-free extracts derived from ascites tumor cells and properly preincubated have an extremely low background. This makes the analysis of exogenous mRNA-directed in vitro products much easier; this situation can be contrasted to the reticulocyte cell-free system, where the background is high and the response to exogenous mRNA is poor [17]. When the system is supplemented with homologous tRNA, its activity is substantially improved and is as efficient as that of systems derived from E. coli [18].

Because the ascites cells have a dual role in providing an active cell-free system and serving as a host for viral replication, the system permits one to correlate the translation of the EMC message in vitro with the viral-directed proteins synthesized in infected ascites cells. For example, it has been proposed that the in vivo synthesis of EMC proteins proceeds through high molecular weight precursors processed into smaller capsid and noncapsid proteins [19]. This problem should be amenable to study in the cell-free system in which the fate of the high molecular weight products synthesized in response to the EMC message can be readily followed.

The system may be useful for the translation of other mRNAs. Evidence for this are the recent observations that globin mRNA derived from rabbit [20-27], human [23], and duck [24] reticulocytes is translated very efficiently in the ascites cell-free system, and that the product synthesized corresponds to authentic rabbit globin. In addition, bovine lens [25] and immunoglobulin light-chain mRNAs [26, 27] are translated efficiently and accurately in this cell-free system.

Potentially, the ascites cell-free system can be used to study certain aspects of gene expression in animal cells. For example, a model that implicates changes in the levels of mRNA for a given gene may be tested by isolation of the specific messenger RNA fraction and quantitation of the message by assay of its product in vitro. Ideally, the product to be studied should not be synthesized by the ascites cells, e.g., (hemo)globin. The importance of this point is that it would be much easier to detect the synthesis of a specific product programmed by an added mRNA in an in vitro system derived from a cell line that does not normally synthesize that product. An inherent assumption to the above approach is that cell-free extracts prepared from ascites cells translate all mammalian mRNAs efficiently and accurately. Also, it is important to keep in mind that mRNAs from differentiated tissues may require corresponding tissue-specific factors in order to be translated in a heterologous system [29].

ACKNOWLEDGMENTS

We are grateful to Dr. Philip Leder for his role in the initiation of these studies and for his active participation throughout the course of the investigation, and for critically reviewing this manuscript. In addition, we are indebted to Dr. A. T. H. Burness for supplying us with Krebs II ascites tumor cells and encephalomyocarditis virus, to Drs.E. M. Martin and A. T. H. Burness and M. Mathews for helpful suggestions and for making preprints of their work available to us.

I. B. was a postdoctoral fellow of the American Cancer
Society (PF-619).

APPENDIX

The formulas for the various buffers and special reagents
discussed in the text follow:

1. Buffer A,

   40.0 g NaCl

   1.0 g KCl

   5.75 g Na_2HPO_4

   1.0 g KH_2PO_4

   5000 ml distilled water

Before use penicillin and streptomycin are added to a
final concentration of 100 units/ml and 60 µg/ml, respectively.

2. Buffer B,

   35 mM Tris·HCl (pH 7.5)

   140 mM NaCl

3. Buffer C,

   10 mM Tris·HCl (pH 7.5)

   10 mM KCl

   1.5 mM magnesium $(acetate)_2$

4. Buffer D,

   30 mM Tris·HCl (pH 7.5)

125 mM KCl

5 mM Mg (acetate)$_2$

7 mM 2-mercaptoethanol

5. Buffer E, <u>Earle's medium (without sodium bicarbonate)</u>

Solution A

68.0 g NaCl

4.0 g KCl

2.0 g MgSO$_4 \cdot$7H$_2$O

0.1 g phenol red

400 ml distilled water

Solution B

1.40 g NaH$_2$PO$_4 \cdot$2H$_2$O

10.0 g glucose

400 ml distilled water

Solution C

2.0 g CaCl$_2$ in 200 ml distilled water

Fresh Earle's medium is prepared as needed by mixing 4 volumes of A, 4 volumes of B, and 2 volumes of C with 90 volumes of distilled water. Penicillin and streptomycin are also added. This solution can be purchased commercially in a 10-times concentrated form.

6. Buffer F, <u>PP-8 buffer</u> (Ref.[16] and E. M. Martin, personnal communication)

Solution A (0.2 M potassium phosphate, pH 7.8)–34.0 g of K$_2$HPO$_4$ dissolved in water, adjusted to pH 7.8 with concentrated phosphoric acid.

Solution B (0.4 M pyrophosphate, pH 8.0) - 178.4 g $Na_4P_2O_7 \cdot 10 \ H_2O$ dissolved in water, adjusted to pH 8.0 with HCl and made up to 1 liter.

To prepare the buffer, dissolve 11.7 g NaCl in 1 liter of Solution A and add 1 liter of Solution B. The resulting solution then contains 0.1 M phosphate, 0.2 M pyrophosphate, and 0.1 M NaCl at pH 8.0. When this solution is stored at 5°C, pyrophosphate may precipitate out of solution. In this case, the solution is decanted before use.

7. Cesium chloride solutions

For density of 1.39:

Dissolve 0.54 g of CsCl into 0.85 ml of water

For density of 1.34:

Dissolve 0.46 g of CsCl into 0.87 ml of water

Both solutions can be scaled up as needed

8. Hemagglutination equipment

"Microtiter" plates and "microdilutors" are trade names and are sold by various biological supply houses, e.g., Flow Laboratories, Rockville, Md., and Microbiological Associates, Bethesda, Md.

Sheep erythrocytes are also available from these companies.

REFERENCES

[1] I. M. Kerr, E. M. Martin, M. G. Hamilton and T. S. Work, Cold Spring Harbor Symp. Quant. Biol., 27, 259 (1962).

[2] I. M. Kerr, N. Cohen, and T. S. Work, Biochem. J., 98, 826 (1966).

[3] M. B. Mathews, and A. Korner, Eur. J. Biochem., 17, 328 (1966).

[4] A. E. Smith, K. A. Marcker, and M. B. Mathews, Nature, 225, 184 (1970).

[5] I. Boime, H. Aviv, and P. Leder, Biochem. Biophys. Res. Commun., 45, 788 (1971).

[6] P. Dobos, I. M. Kerr, and E. M. Martin, J. Virol, 8, 491 (1971).

[7] E. M. Martin, J. Malec, S. Sved, and T. S. Work, Biochem. J., 80, 585 (1961).

[8] H. Aviv, I. Boime, and P. Leder, Proc. Nat. Acad. Sci. 68, 2303 (1971).

[9] G. Klein and E. Klein, Ann. N.Y. Acad. Sci.. 63, Art. 5, 895 (1956).

[10] H. Goldie, Ann. N.Y. Acad. Sci., 63, Art. 5, 711 (1956).

[11] G. von Ehrenstein, in Methods in Enzymology (1. Grossman and K. Moldave, eds), Vol. 12, Academic Press, New York, 1968, p. 588.

[12] A. T. H. Burness, A. D. Vizoso, and F. W. Clothier, Nature, 197, 1177 (1963).

[13] P. H. Hofschneider and P. Hausen, in Molecular Basis of Virology, ed. (H. Fraenkel-Conrat, ed.), ACS Monograph 164, Reinhold, New York, 1968, p. 176.

[14] A. J. D. Bellett and A. T. H. Burness, J. Gen Microbiol., 30, 131 (1963).

[15] C. W. Jungeblut, in Handbuch der Virus Forschung (C. Hallaver and K. F. Meyer, eds.), Vol. 4, Springer-Verlag, Berlin, 1958, p. 459.

[16] A. T. H. Burness, J. Gen Virol., 5, 291 (1969).

[17] R. E. Lockard and J. B. Lingrel, Biochem. Biophys. Res. Commun., 37, 204 (1969).

[18] D. Nathans, in Methods in Enzymology (L. Grossman and K. Moldave, eds.), Vol. 12, Academic Press, New York, 1969, p. 787.

[19] B. E. Butterworth, L. Hall, C. Stoltzfus, and R. R. Rueckert, Proc. Nat. Acad. Sci., 68, 3083 (1971).

[20] D. Housman, R. Pemberton, and R. Taber, Proc. Nat. Acad. Sci., 68, 2716 (1971).

[21] M. B. Mathews, M. Osborn, and J. R. Lingrel, Nat. New Biol., 233, 206 (1971).

[22] H. Aviv, and P. Leder, Proc. Nat. Acad. Sci., in press.

[23] S. Metafora, M. Terada, L. W. Dow, P. A. Marks, and A. Banks, Proc. Nat. Acad. Sci., 69, 1299 (1972).

[24] H. Aviv, S. Packman, and P. Leder (unpublished).

[25] M. B. Mathews, M. Osborn, A. J. M. Berns, and H. Bloewendal, Nature New Biology, 236, 5 (1972).

[26] G. G. Brownlee, T. M. Harrison, M. B. Mathews, and C. Milstein, FEBS Letters, 23, 242 (1972).

[27]   D. Swan, H. Aviv, and P. Leder, <u>Proc. Nat. Acad. Sci.</u>, <u>69</u>, (1972).

[28]   I. M. Kerr and E. M. Martin, <u>J. Virol</u>, <u>9</u>, 559 (1972).

[29]   S. M. Heywood, <u>Proc. Nat., Acad. Sci.</u>, <u>62</u>, 1782 (1970).

Chapter 9

PREPARATION OF 5 S RNA

Robert Röschenthaler and Peter Herrlich

Institut für Hygiene und Medizinische
Mikrobiologie der Technischen Hochschule München
Munich, West Germany
and
Max-Planck-Institut für Molekulare Genetik
Berlin-Dahlem, West Germany

I. INTRODUCTION

A. Discovery

The presence in ribosomes of low-molecular-weight RNA was
first described in 1959 [1—4]. Some properties of this RNA
seemed to differ from normal tRNA [5]. A clear-cut distinction
between this RNA and tRNA, however, was first made by Rosset
and Monier in 1963 [6] with RNA extracted from Escherichia coli.
It was found that the sedimentation coefficient in a sucrose
gradient was slightly higher than that of tRNA (molecular
weight 40,000 [6—8]); therefore Rosset and Monier designated
this kind of RNA 5 S RNA. A series of other properties was
reported by these authors which clearly distinguished 5 S RNA
from tRNA, as well as from 16 and 23 S rRNA. The 5 S RNA was
found to be attached to the 50 S ribosomal subunits and had
a chain length of more than 100 nucleotides [7]. Comb and
Katz [8] independently discovered an RNA component from ribosomes
of the aquatic fungus Blastocladiella emersonii which proved to
have properties similar to the 5 S RNA described by Marcot-Queiroz
and Monier [10]. Since it was found that his RNA was neither a
precursor of tRNA nor a degradation product of high-molecular-
weight rRNA, but rather a distinct species of rRNA, the
designation 5 S RNA was also adopted for this RNA component.

Since these studies, 5 S RNA has been discovered in
various organisms, prokaryotic as well as eukaryotic, including
bacteria [11], fungi [8—10], plants [12], echinoderms [13],
insects [14], and vertebrates [15—17]. 5 S RNA therefore
appears to be a universal constituent of the ribosomes of all
living cells.

B. Properties

The properties of 5 S RNA have recently been reviewed [18].
The primary structures of 5 S RNA from E. coli and KB cells
were determined [19—21] and were found to be different. The
length of the molecules, however, is identical, both containing
120 nucleotides. 5 S RNA contains no methylated nucleotides
[7,9,12,22]. It has a pronounced secondary structure. Measure-
ments of the hypochromicity of E. coli 5 S RNA indicate about
60% base pairings of the nucleotides [23]. Optical rotatory
dispersion suggests that 67—82% of the molecule is base paired
and that 60—70% of these are GC pairs [23].

5 S RNA is tightly bound to the 50 S ribosomal subunits;
each subunit contains one molecule. It cannot be released
from the particle in 10^{-4} M Mg^{2+} solutions [24], whereas tRNA
is detached under these conditions [9].

Treatment of 50 S subunits with EDTA [25] results in
unfolding of the ribosomes without loss of proteins. These
particles sediment at approximately 21 and 36 S [25] and lose
their 5 S RNA [24,26]. Also, dialysis against 0.5 M NH_4Cl in
the absence of Mg^{2+} results in dissociation of 5 S RNA from the

50 S subunits [27]. If 50 S ribosomes are treated with 2 M LiCl,
about half of the proteins, together with the 5 S RNA, are
removed [24].

5 S RNA has been isolated in two forms, A and B, which can
be separated by column chromatography [28]. The A form resembles
native 5 S RNA; the B form is obtained by denaturation with heat
or urea. Isolated B form can be converted into A form by heating in the presence of Mg^{2+} or in media of high ionic strength.
This conversion reflects a change in conformation of the
molecule [28].

Only native 5 S RNA or the A form can be reattached to
LiCl-stripped particles in the presence of the split proteins
[24,28]. The B form has no affinity for such particles nor for
the split proteins. Thus it appears that 5 S RNA plays its
biological role, whatever this may be, when in a certain molecular
conformation.

There seems to be no pool of free 5 S RNA, but there is a
small pool of immediate precursors of 5 S RNA in the cell sap.
These immediate precursors of 5 S RNA are present in normal
exponentially growing E. coli and after chloramphenicol treatment of the bacteria [29,30]. They are longer at the 5' end
than 5 S RNA by one of three nucleotides [30]. Also, during
amino acid starvation of an RC^{rel} strain of E. coli ($RC^{relaxed}$)
and $RC^{str(ingent)}$ are designations of bacterial phenotype in
which the synthesis of RNA is not or is obligatorily coupled to
RNA synthesis, respectively a precursor 5 S RNA that was

slightly longer than the mature molecule was found outside the
ribosomes [31]. It is assumed that this precursor 5 S RNA is
shortened when it is incorporated into ribosomal precursor
particles. 32 and 43 S premature ribosomal subunits from E.
coli do contain a 5 S-like RNA [31]. In eukaryotes (HeLa cells)
newly synthesized 5 S RNA appears in the cytoplasm later than
newly formed 28 S RNA, suggesting the occurrence of a nuclear
pool of 20—30% of the total cellular 5 S RNA. The earlier
suggestion that in eukaryotic cells 5 S RNA derives from the
45 S precursor of higher-molecular-weight RNA is no longer
accepted [32]. The presence of 5'-di- and triphosphate groups
in 5 S molecules indicates that 5 S RNA in HeLa cells does not
derive from a precursor containing an extra sequence at its 5'
end as is the case in E. coli [33].

Unfortunately, no mutants defective in 5 S RNA synthesis
have been described, and the 5 S RNA gene or genes have not
been mapped. There is conclusive evidence, however, that 5 S
RNA in E. coli is transcribed in a large transcriptional unit
that is 13 to 30 times the length of the 5 S RNA gene [34], and
that 5 S RNA is formed by posttranscriptional cleavage [35] linked
to 5 S RNA during synthesis. Such a linkage was proposed for
Bacillus subtilis on the basis of several kinetic studies
[36,37] and actinomycin titrations [38]. There seems to be no
linkage in eukaryotes [39,40].

The biological function of 5 S RNA is still obscure. It
was suggested [26] that it may have a role in the specific

binding of tRNA to ribosomes. A stimulatory effect on the
incorporation of amino acids directed by MS2 RNA, but not by
poly U, has been reported [41]. The reason for this stimulation
is not yet known.

II. PREPARATION OF 5 S RNA FROM RIBOSOMAL PARTICLES

The early isolations of 5 S RNA [6,7] were made by
chromatography of whole-cell phenol extracts on methylated
serum albumin—kieselguhr columns (MAK columns [42]). The
resolution of the 5 S RNA from tRNA by this method, however,
was not very satisfactory. Therefore a method was introduced
for extracting 5 S RNA from ribosomal particles from which tRNA
had been previously removed [9]. Extraction of these particles
yielded 5 S RNA and high-molecular-weight rRNA. High-molecular-
weight RNA, however, can be easily separated from 5 S RNA,
either by MAK column chromatography or by specifically precipitating the high-molecular-weight RNA with 2 M NaCl. 5 S RNA
is soluble at this concentration of NaCl.

This method yields pure preparations of 5 S RNA. As only
the particles are extracted, it cannot give information concerning problems dealing with free pools of 5 S RNA or their precursors.

A typical procedure for a preparation of 5 S RNA by this
method is as follows. Escherichia coli cells are grown in
either defined or peptone media at 37°C with aeration and
agitation. The cells are poured onto frozen TM-2 buffer

[10^{-2} M tris—HCl (pH 7.4), 10^{-2} M $MgCl_2$], centrifuged at
15,000 X g for 5 min, and washed twice with TM-2 buffer.

About 6 g of the bacterial paste is ground with twice its
weight of alumina (Alcoa Company, bacteriological grade) for
10—20 min in an ice bath (see Vol I, Chapter 7). At the end
of the grinding, an amount of TM-2 buffer is added that makes
the paste just fluid enough to poured into centrifuge tubes.
After the alumina and cell debris are removed by centrifugation
(5 min at 15,000 X g), deoxycholate (5% solution) is added to
the supernatant to give a concentration of 0.2%. The supernatant is again centrifuged (at 30,000 X g for 30 min) and the
supernatant is carefully decanted off and centrifuged for 2 hr
at 105,000 X g. The ribosome pellet is washed once with TM-4
buffer (containing 10^{-4} M $MgCl_2$). The pellet is resuspended
in the same buffer and dialyzed overnight against this buffer
at 4°C. Precipitates, if formed, are removed by low-speed
centrifugation. The ribosomal subunits are sedimented by
centirifugation at 105,000 X g for another 4 hr; the supernatant
is discarded, and the sides of the tubes and the surface of the
pellet are washed once with TM-4 buffer. The pellet is dissolved
in about 10 vol of 0.1 M NaCl, 0.1% sodium dodecyl sulfate
(SDS), and the solution is extracted by shaking with an equal
volume of TM-2-saturated phenol for about 10 min. The phenol
phase is then separated from the aqueous phase by low-speed
centrifugation. The aqueous phase is recovered by carefully

pipetting it from the phenol phase. The phenol layer is twice
extracted with an equal amount of TM-2 buffer, and the aqueous
phase are combined and centrifuged again at 30,000 X g for 30
min to clarify the extract of residual protein. To the aqueous
extract 10% of its volume of 2 M sodium acetate (pH 5.2) is added
and the RNA is precipitated by the addition of 2 vol of ethanol.
The mixture is kept at -20°C overnight; the precipitate is then
collected by centrifugation (10,000 X g for 5 min) and the RNA
is dissolved in 0.2 M NaCl. This fraction contains mainly 5 S
RNA and high-molecular-weight rRNA.

At this stage there are two choices for removing the high-
molecular-weight RNA: precipitation by high salt concentrations,
or separation by column chromatography, for example, MAK columns.

A. Salting Out of the High-Molecular-Weight RNA

The RNA solution is made 2 M by the addition of 4 M NaCl.
At this concentration high-molecular-weight RNA, especially
23 S RNA, is precipitated rather rapidly. However, in order to
obtain a nearly quantitative precipitation, the solution is
allowed to stand for at least 12 hr at -10°C. The precipitate
is then removed by centrifugation and the pellet is washed twice
with ice-cold 1 M NaCl solution. The supernatants from the
washings are combined with the first supernatant and diluted
at least to 1 M NaCl with water. The 5 S RNA is precipitated
by ethanol as above. The pellet is dissolved in a small volume
of TM-2 buffer; it can be kept in the frozen state for more than

a year and will still give the same chromatographic patterns.
Difficulties may arise if the 2 M NaCl solution is not diluted
before the precipitation with ethanol, as a result of crystal-
lization of salt in presence of ethanol. Another possibility
is the use of a solution of 2 M LiCl instead of NaCl. LiCl does
not crystallize in ethanol at this concentration. It was
observed, however, that some 5 S RNA precipitates at this LiCl
concentration [9].

B. Removing the High-Molecular-Weight RNA
by MAK Column Chromatography

Since only 3% of the total rRNA is 5 S RNA, MAK columns,
which have a capacity of a few milligrams of RNA, are not
efficient when the total RNA is applied. Therefore, when
milligram amounts of 5 S RNA must be prepared, it is reasonable
to use the salting-out procedure, even in a shortened form
(only a few hours of precipitation in 2 M NaCl), before
separation on MAK columns (Figure 1).

MAK columns can be prepared as follows [42]:

(1) Kieselguhr: Hyflo-Super-Cel (Johns-Mansville) is a
diatomaceous earth with prevailing particle size from 5 to 25 μ.

(2) Bovine serum albumin: Fraction-V powder (Armour
Laboratories, Nutritional Biochemicals, Sigma, Calbiochem).

(3) Methylated bovine serum albumin (MBSA): This is
prepared by dissolving 10 g of serum albumin in 1 liter of
absolute methyl alcohol and adding, dropwise with agitation,

8.5 ml of 12 N HCl. During the addition the serum albumin

dissolves completely. The mixture is allowed to stand in the

dark at room temperature for 5 days or at 37°C for 2 days.

The protein precipitates again. The precipitate is collected

by centrifugation at 12,000 X g (10 min) and is washed twice

with methanol and twice with ether. The paste is then ground

in a mortar until it is dry. The powder is stored over KOH

under reduced pressure at 4°C. A 1% solution in water is

stable in a refrigerator for several months. Difficulties

may arise if MBSA is not washed well enough or if it is not

dried well enough by the ether treatment. Then the solubility

in water may not be complete.

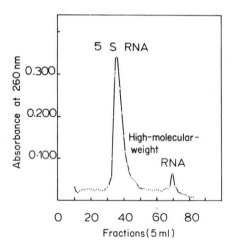

Fig. 1. Separation on a MAK column of 5 S RNA, from contaminating high-molecular-weight RNA, after the salting-out
procedure.

(4) Protein-coated Super-Cel: 20 g of Super-Cel are suspended in 100 ml of 0.1 M buffered NaCl solution, boiled to expel air, and cooled; 5 ml of 1% MBSA are added with stirring, together with 20 ml of 0.1 M NaCl. The suspension is transferred into a column of 2 X 18 cm; a shallow layer of cellulose powder is put on the sintered-glass disc. The Super-Cel is then washed with 200—300 ml of 0.4 M buffered NaCl solution and suspended. The suspension can be kept for about 2 months in the refrigerator

(5) Salt solutions: NaCl solutions are prepared in 0.05 M phosphate buffer to give a final pH of 6.7. In this chapter only the NaCl concentration is mentioned.

(6) Preparation of the column: A suspension of cellulose powder (Whatman standard) is put on the sintered-glass disc of a 2 X 8 cm column to give a layer of about 2 mm. The next layer on the column is made by addition of a suspension consisting of 8 g of deaerated Super-Cel in 40 ml of 0.1 M NaCl and 2 ml of 1% MBSA. A suspension containing 6 g of Super-Cel in 40 ml of 0.4 M NaCl and 10 ml of MBSA-coated Super-Cel is then layered on top of the Super-Cel. Onto this layer a protective layer is added that consists of 1 g of Super-Cel in 10 ml of 0.4 M NaCl solution. The salt solutions are allowed to drain between each layer, but care must be taken that no air enters the column. After this procedure the column is washed with about 200 ml of 0.2 M NaCl solution (the concentration of the starting buffer). The RNA (1—4 mg) is then dissolved in

50—60 ml of the starting buffer and added onto the column; the
column is washed again with about 100 ml of the same buffer.
A gradient consisting of 250 ml of 0.2 M NaCl and 250 ml of
1—1.5 M NaCl is then applied at room temperature, the flow
rate being about 60—100 ml/hr. The fraction size is usually
2—10 ml. The sequence in which the nucleic acids are eluted
from such a column is: tRNA, 5 S RNA, DNA, 16 and 23 S RNA.
If no tRNA or DNA is present in the RNA preparations, 5 S RNA
is well separated from the high-molecular-weight RNAs. The
columns are usually used only once since their resolution in a
second run is poorer. These columns are also very useful for
testing the purity of 5 S RNA in relation to small amounts of
contaminating DNA or high-molecular-weight RNA. RNase is more
strongly retained on MAK columns than DNA [43]. This property
should yield 5 S RNA free of traces of RNase. As mentioned
initially, it is possible to separate the denatured (B) form
of 5 S RNA on MAK columns [28]. The B form is eluted after
the native (or A) form. The denatured form was found when
MAK chromatography was applied to native 5 S RNA at 37°C and
therefore suggested that the MAK column causes a change in
conformation at this temperature [44]. The immediate precursor
of 5 S RNA is eluted from MAK columns behind native, mature
5 S RNA [39].

III.   PREPARATION OF 5 S RNA FROM WHOLE-CELL EXTRACTS
        Since efficient MAK column chromatography of 5 S RNA

implies tRNA-free RNA preparations, and because of the limited
separatory capacity of MAK columns, this method is restricted
to RNA samples that have already been partially purified.
Some methods that can be applied to the total, RNA-containing
extract of cells are described below.

A. Separation of 5 S RNA on Sephadex G-75

Chromatography on Sephadex G-100 or G-75 (Pharmacia, Uppsala,
Sweden) was used by several investigators [11,22,28,33,45].
The elution buffer used varied, but all have in common the use
of rather long columns of Sephadex. The lengths varied from
about 100 to 280 cm; the longer the columns the better the
resolution of 5 S RNA from tRNA. Unfortunately, the flow rate
decreases with length.

Preparation of a G-75 Sephadex Column [45]

Sephadex is swollen in an excess of 5 mM potassium acetate
for 2 days at room temperature. The suspension is agitated
from time to time and the fine grains and slowly sedimenting
particles are decanted. The slurry is then deaerated with a
suction pump until no more air bubbles develop and cooled to
$4°C$. To the bottom of the glass column (with a size of 2 X
150—200 cm), a small plug of glass wool is added and, above
this, a layer of glass beads 2—3 cm high is added (1-mm
diameter). The slurry of Sephadex is poured onto this support
and the Sephadex particles are allowed to sediment by gravity.
During this time the column may be drained, but care must be

taken that it does not become dry. More slurry should be added
before a clear column of buffer develops above the Sephadex.
When the column is replenished and settled, a small disc of
filter paper and a small amount of glass wool are placed on the
top to protect the surface. The packing of the column can be
tested by charging it with a small amount of blue Dextran 2000
(Pharmacia). Elution of this material with 5 mM potassium
acetate (pH 4.2) gives a rather sharp band, the elution of which
can be observed visually. The RNA sample is applied in a
minimal volume; 100—300 mg of RNA can be applied to columns of
the above dimension. The column is run at a temperature of 4°C.
The flow is established by gravity and the reservoir contains
600 ml of buffer. (Never apply air pressure to Sephadex columns.
To avoid the high water pressure in long columns, it may even
be advisable to use shorter ones in series.) A flow rate of
about 10 ml/hr (2- to 10-ml fractions) is usually obtained,
which means that for complete elution about 2 days are necessary.

   Sephadex allows the separation of molecules according to
size. The first peak contains excluded high-molecular-weight
RNA and aggregated denatured tRNA [45] (RNA that is too large
to enter the Sephadex network), the second peak is 5 S RNA, and
the third peak consists of tRNA (Figure 2). (With very slow
flow rates, a fourth peak can be obtained which also consists
of tRNA.)

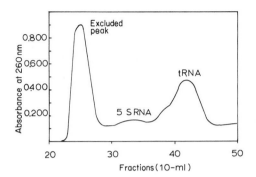

Fig. 2.   Separation of 5 S RNA from a whole-cell extract
on Sephadex G-75.   Average separation achieved.

Alternatively, Sephadex G-100 columns (3 X 280 cm) eluted
with a buffer of 0.25 M NaCl—10 mM sodium acetate (pH 5.0)—
1% (v/v) methanol, or standard saline citrate buffer (0.15 M
NaCl, 0.015 M sodium citrate), were described [28,29].

B.   Gel Filtration of Crude Extracts

An interesting technique of Sephadex gel filtration uses
crude cell extracts that are not deproteinized but contain
small amounts of SDS [46].   Only after the separation are the
samples deproteinized by phenol extraction.   Even rather short
columns (60 X 2.5 cm) give good resolutions.

Preparation of the Extracts

Crude extracts from E. coli K12 cells are obtained by
disruption in a French pressure cell (15,000—17,000 psi).

After centrifugation at 30,000 X g for 30 min, either ribosomes
or the crude extract, with a final concentration of 0.25% SDS
for clearing, are prepared (SDS 99% pure, Pfaltz and Bauer,
New York). Ribosome suspensions require a final concentration
of 0.5% SDS to become clear. The Sephadex G-75 is swollen
before use in 0.1 M ammonium acetate (pH 5.2)—0.1% SDS. This
buffer is also used for the elution of the column when crude
extracts are employed. For ribosome suspensions the elution
buffer consists of 0.01 M tris, 0.1 M NaCl, and 0.25% (pH 7.3).
The columns are run at room temperature to avoid precipitation
of the SDS. The sequence of the peaks resembles that obtained
when deproteinized RNA samples are employed.

C. Separation of 5 S RNA from tRNA by DEAE-Cellulose
Chromatography

Comb and Zehavi-Willner [9] separated tRNA from 5 S RNA on
DEAE-cellulose at 80°C. At this temperature random coil forms
are separated. At room temperature no separation is achieved.

Preparation

DEAE-cellulose (capacity 0.65 meq/g) is first washed with
an excess of 0.1 M NaOH, then with 0.1 M HCl, and finally with
0.1 M NaCl until neutral. The suspension of DEAE-cellulose in
0.1 M NaCl is carefully deaerated, for example, with the use of
a suction pump, and brought to 80°C. The glass column has a
water jacket that is connected to a water bath at 80°C. The hot
suspension of DEAE-cellulose is poured into the column (1 X 10 cm)

and is washed at this temperature with 1.5 M NaCl——0.02 M

tris—HCl (pH 7.3), followed by 0.3 M NaCl——0.02 M tris—HCl

(pH 7.3). Care must be taken that all buffers used are well

deaerated and brought to 80°C before use. Otherwise the column

will be ruined by enclosed air bubbles. About 10—20 A_{260}

units of RNA are applied. The column is eluted with a linear

gradient of 150-ml each of 0.3 M NaCl——0.02 M tris—HCl (pH 7.3)

and 1.5 M NaCl—0.02 M tris HCl (pH 7.3). The gradient solutions

may be maintained at room temperature but should be deaerated.

A flow rate of about 10 ml/hr and 5-ml fractions are convenient.

D. Separation of 5 S RNA from a Whole-Cell RNA Extract
by Polyarginine—Kieselguhr Column Chromatography

A fast and convenient separation of 5 S RNA from other

nucleic acids was possible by chromatography on columns of

kieselguhr with polyamino acids (polyornithine, polylysine,

polyarginine) [47]. The best separation of 5 S RNA was achieved

on polyarginine columns.

Preparation

Eight grams of Hyflo-Super-Cel (Johns-Mansville) are boiled

and cooled in 40 ml of 0.4 M NaCl—0.06 M sodium phosphate (pH

6.7), and a solution containing 6 mg of polyarginine is added.

The polyarginine (Yeda-Miles Inc., Rehovot, Israel) is dissolved

in 0.03 M HCl at a concentration of 2 mg/ml. After it is

thoroughly mixed, the slurry is poured onto a thin layer of

cellulose powder resting on a fritted disc in a glass column

2 X 15 cm. After the slurry has settled, a protective layer
of Hyflo-Super-Cel (in a suspension of 2 g/10 ml starting
buffer, boiled and cooled) is added to the top of the column.
The column is washed with 150 ml of starting buffer, the RNA
sample (50—100 A_{260} units) is applied in a volume of 10—20 ml
of starting buffer, and an elution gradient is connected to
the column. The gradient consists of 1.0 and 5.0 M NaCl (300
ml each) buffered with 0.06 M sodium phosphate (pH 6.7) and
with 0.1 M tris—HCl (pH 8.9), respectively.

The 5 S RNA peak is eluted after the tRNA peak and long
before the DNA peak; the high-molecular-weight RNA is not or is
only partially eluted from the column (Figure 3). It can be
removed from the column with 5 M NaCl solution at pH 10. If
one wishes to avoid extensive washing with 5 M NaCl solution
(pH 10) before reusing the column, the high-molecular-weight
RNA should be salted out with 2 M NaCl before application.

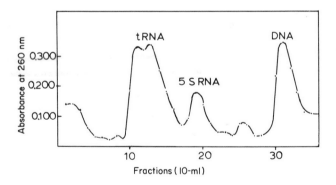

Fig. 3.   Separation of 5 S RNA from a whole-cell extract
on a polyarginine column.

The columns can be reused. They are first washed with
5.0 M NaCl solution (150 ml) and then equilibrated with starting
buffer (1.0 M NaCl). The columns are used at room temperature.
The elution lasts about 2 hr. They are therefore useful for
rapid analytical work.

E. Separation of 5 S RNA from Nuclear Low-Molecular-Weight RNA Species by DEAE-Sephadex Chromatography

Moriyama et al. [48] described the fractionation by DEAE-
Sephadex chromatography of 4—7 S RNA species, observed in the
nuclei of mammalian cells, at pH 3.4, 5.1, and 7.6. Excellent
separations of 5 S RNA and other low-molecular-weight RNAs were
achieved in quantities of up to 30 mg of RNA. Their purity was
determined by acrylamide gel electrophoresis. Three different
peaks of 5 S RNA were shown.

Preparation of the Columns

DEAE-Sephadex A-50 (Pharmacia, Uppsala, Sweden), with a
capacity of 3.5 meq/g, is swollen overnight in water and fine
particles that do not settle in 15 min are decanted. To obtain
good flow rates, this procedure is repeated several times. The
sedimented DEAE-Sephadex is then washed with an excess of 0.1 M
HCl, followed by water, 0.1 M NaOH, water, 1.0 M NaCl and,
finally, water. The material is then suspended in the appropriate
starting buffer. After equilibration with the starting buffer,
the column (0.9 X 120 cm) is packed under gravity and washed

with 300 ml of the initial buffer. The RNA sample, at a
concentration of 10 mg/ml in water, is applied to the column
and eluted with a linear gradient of 2 X 500 ml of buffered
NaCl solution. The concentration of this solution depends on
the pH of the buffer with which it is prepared. For chromatography at different pH, the solutions shown are used.

pH	NaCl gradient	Made up in buffer solution
3.4	0.58—0.72 M	0.02 M Citrate buffer (pH 3.4)
5.1	0.60—0.72 M	0.02 M Acetate buffer (pH 5.1)
7.6	0.45—0.60 M	0.02 M Tris—HCl buffer (pH 7.6)

Fraction of the 5 S RNA is obtained at pH 5.1. The flow
rate is maintained at 10 ml/hr; higher flow rates give poorer
resolution. Fractions of 3.2 ml were collected in the example
mentioned.

F. Polyacrylamide Gel Electrophoresis

Polyacrylamide is a supporting gel with a definite average
pore size. The pore size varies with the initial concentration
of the monomers polymerized. The electropheretic mobility of
RNA is inversely related to its sedimentation coefficient (this
is true for intermediate pore sizes that are in the range of
the size of the molecules to be separated). Differences in the
ratio of charge to mass for different RNA molecules are likely
to be too small to affect electrophoretic mobilities. The

secondary structure of RNA, however, may influence its migra-
tion. Resolution of RNAs depend, therefore, on molecular
filtration (that is, on the pore size of the gel versus the
size of the molecules).

For the separation of 4 and 5 S RNA, higher gel concentrations are most suitable, that is, above 7% acrylamide. The
separation of 4 and 5 S RNA by this method was first described
by J. Hindley (cited in Ref. 49) and by Cannon and Richards
[50]. Microsomal 5 S RNA from mammalian cells was separated
from other RNA species by Peacock and Dingman [51].

Technique [51—53]

Short cylindrical gels are prepared by polymerizing
acrylamide, together with a small proportion of crosslinking
agent (N,N'-methylenebisacrylamide), in Plexiglas (Perspex)
tubes (preferred to glass, which is also used). For good
resolution, especially of larger RNA molecules, it is advisable
to recrystallize both monomers before use (see Ref. 53).

The monomers are stable in aqueous solution at 4°C in the
dark. Both acrylamide and bisacrylamide are very poisonous and
all precautions should be taken in order not to inhale their
vapor from solutions or crystals and to prevent skin contact.

In the literature, usually only the concentration of
acrylamide is given (2—15%). Final bisacrylamide concentrations
are 2.5—5% that of the acrylamide. For example, to separate
5 and 4 S RNA on a 7.5% gel, both monomers are mixed at a final

concentration of 7.5% acrylamide and 0.25% bisacrylamide. The
solution should be degassed under reduced pressure and tris
acetate (pH 8.3) added to a final concentration of 0.04 M. To
polymerize the monomers, 0.033 ml of N,N,N'N'-tetramethylethy-
lenediamine and 0.33 ml of 10% (w/v) ammonium persulfphate are
added per gram of acrylamide present. The solution is mixed
without aeration and rapidly pipetted into the tubes. Water is
layered on top to ensure a flat surface. Because it is desirable
that the RNAs enter the gel in a sharp zone, the water can be
replaced later by a large-pore gel where the RNAs can migrate
almost without hindrance (for the exact method see Refs. 54 and
55). Polymerization of the gel requires 2—10 min, depending
on room temperature. The RNA sample is then applied (5—100
µg/gel in 5% sucrose). Both ends of the gels are connected
to buffer reservoirs and, for the tris acetate buffer mentioned,
10 V/cm at 5 mA per gel is applied for about 1—3 hr.

G. Polyacrylamide Gel Slab Electrophoresis

Flat-slab gel electrophoresis is a preparative method
developed by Sanger and Brownlee. Adams et al. [56] used flat
slabs for separating oligonucleotides that arose from the
partial digestion of R17 RNA. Jordan applied this technique
successfully for the preparative purification of 5 S RNA [57].

1. Technique

Two glass plates (40 X 20 X 0.4 cm) are thoroughly cleaned.
Along their long sides the glass plates are separated from each

other by two vaseline-coated spacers (40 X 2 X 0.3 cm) so that
a flat space is formed between the glass plates. The assembly
is kept together by strong spring clips. The bottom is sealed
with Plasticine. A few milliliters of gel are poured in and
polymerized in order to detect eventual leakage. Then the
space is filled completely, to the top, where sample wells are
formed by pushing a suitable slotted strip of Perspex into the
top of the gel (15 X 3 X 0.2 cm). After gelation the well
former and Plasticine are removed. The gel is made from 12.1%
acrylamide, for example, plus 0.4% bisacrylamide in 0.04 M tris
acetate (pH 8.3); 1/1000 vol of N,N,N'N'-tetramethylethylene-
diamine and 3/1000 vol of freshly dissolved 10% (w/w) ammonium
persulfphate are added. Polymerization requires 5—15 min at
room temperature. The RNA samples are mixed with 1/3—1 vol
of 50% sucrose and placed into the wells. The bottom of the
gel is placed in a container with the same buffer as above; the
top is connected to a similar container by a strip of Whatman
3MM paper. Electrophoresis is performed at 400 V, 40 mA for
16 hr at 2—4°C.

    With the described method and slab size, 4 and 5 S RNA are
separated by about 3—4 cm (Figure 4). If bromphenol blue (1%)
is used as a marker, 5 S RNA moves with an R_f of about 0.3 with
respect to bromphenol blue.

    When ^{32}P-labeled RNA is used, the RNA can be localized by
placing an x-ray film on top of the gel. The glass plate on
that side is removed, and the film is exposed for a suitable

time.   The film needs no unwrapping and the exposure can be made in daylight.

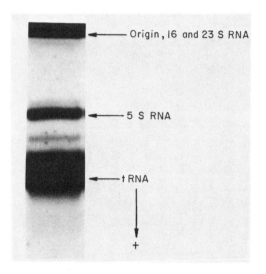

Fig. 4.   Flat-slab gel electrophoresis of a whole-cell extract with phenol.   Whole E. coli cells extracted with phenol give mainly low-molecular-weight RNA.   For the identity of the additional bands, see Ref. 57.   (Courtesy of R. B. Jordan.)

2.   Elution from the Gel

Pieces of gel are placed in a 10-ml syringe without a needle and pushed through to form small pieces or a gel paste; the gel is then filled into another syringe (column like), eluted with MAK buffer (10—50 ml), and the RNA is precipitated.

ACKNOWLEDGMENT

We are grateful to Dr. Jordan for drawing our attention to his method of slab gel electrophoresis and for discussing the details with us, and to Drs. M. Schweiger and T. B. Smith for reading the manuscript.

REFERENCES

[1] T. Hultin and A. Von Der Decken, Exptl. Cell. Res., 16, 444 (1959).

[2] L. Bosch, H. Bloemendal, and L. Sluyser, Biochim. Biophys. Acta, 34, 272 (1959).

[3] M. Takanami, Biochim. Biophys. Acta, 55, 132 (1962).

[4] K. C. Smith, Biochemistry, 1, 866 (1962).

[5] D. Elson, Biochim. Biophys. Acta, 53, 232 (1961).

[6] R. Rosset and R. Monier, Biochim. Biophys. Acta, 68, 653 (1963).

[7] R. Rosset, R. Monier, and J. Julien, Bull. Soc. Chim. Biol., 46, 87 (1964).

[8] D. G. Comb and S. Katz, J. Mol. Biol., 8, 790 (1964).

[9] D. G. Comb and T. Zehavi-Willner, J. Mol. Biol., 23, 441 (1967).

[10] J. Marcot-Queiroz and R. Monier, Bull. Soc. Chim. Biol., 48, 446 (1966).

[11] P. Morell, I. Smith, D. Dubnau, and J. Marmur, Biochemistry, 6, 258 (1967).

[12]   A. K. Chakravorty, Biochim. Biophys. Acta, 179, 67 (1969).

[13]   J. Sy and K. McCarty, Biochim. Biophys. Acta, 199, 86 (1970).

[14]   K. T. Tartof and R. P. Perry, J. Mol. Biol., 51, 171 (1970).

[15]   F. Galibert, C. J. Larsen, J. C. Lelong, and M. Boiron,
Nature, 207, 1039 (1965).

[16]   F. Galibert, C. J. Larsen, J. C. Lelong, and M. Boiron,
J. Mol. Biol., 21, 385 (1966).

[17]   Y. Moriyama, P. Ip, and H. Busch, Biochim. Biophys. Acta,
209, 161 (1970).

[18]   G. Attardi and F. Amaldi, Ann. Rev. Biochem., 39, 183
(1970).

[19]   G. G. Brownlee, F. Sanger, and B. G. Barrell, Nature, 215,
735 (1967).

[20]   G. G. Brownlee, F. Sanger, and B. G. Barrell, J. Mol.
Biol., 34, 379 (1968).

[21]   B. G. Forget and S. M. Weissman, Science, 158, 1695 (1967).

[22]   T. Schleich and J. Goldstein, J. Mol. Biol., 15, 136
(1966).

[23]   B. J. Lewis and T. Doty, Nature, 225, 510 (1970).

[24]   M. Aubert, R. Monier, M. Reynier, and J. F. Scott,
Symposium on Structure and Function of Transfer RNA and 5 S-RNA,
Fourth Meeting of the Federation of European Biochemical Societies,
Oslo 1967, Universitatsforlaget, Oslo, and Academic Press, New
York, 1968.

[25]   R. F. Gesteland, J. Mol. Biol., 18, 356 (1966).

[26] M. A. Q. Siddiqui and K. Hosokawa, <u>Biochem</u>. <u>Biophys</u>. <u>Res</u>. <u>Commun</u>., <u>36</u>, 711 (1969).

[27] M. A. Q. Siddiqui and K. Hosokawa, <u>Biochem</u>. <u>Biophys</u>. <u>Res</u>. <u>Commun</u>., <u>32</u>, 1 (1968).

[28] M. Aubert, J. F. Scott, M. Reynier, and R. Monier, <u>Proc</u>. <u>Nat</u>. <u>Acad</u>. <u>Sci</u>. <u>USA</u>, <u>61</u>, 292 (1968).

[29] F. Galibert, J. C. Lelong, C. J. Larsen, and M. Boiron, <u>Biochim</u>. <u>Biophys</u>. <u>Acta</u>, <u>142</u>, 89 (1967).

[30] B. R. Jordan, J. Feunteun, and R. Monier, <u>J</u>. <u>Mol</u>. <u>Biol</u>., <u>50</u>, 605 (1970).

[31] F. Galibert, M. E. Eladari, A. Hampe, and M. Boiron, <u>European</u> <u>J</u>. <u>Biochem</u>., <u>13</u>, 281 (1970).

[32] D. D. Brown and C. S. Weber, <u>J</u>. <u>Mol</u>. <u>Biol</u>., <u>34</u>, 661 (1968).

[33] L. E. Hatlen, F. Amaldi, and G. Attardi, <u>Biochemistry</u>, <u>8</u>, 4989 (1969).

[34] W. F. Doolittle and N. R. Pace, <u>Nature</u>, <u>228</u>, 125 (1970).

[35] B. Pace, R. L. Peterson, and N. R. Pace, <u>Proc</u>. <u>Nat</u>. <u>Acad</u>. <u>Sci</u>. <u>USA</u>, <u>65</u>, 1097 (1970).

[36] I. Smith, D. Dubnau, P. Morell, and J. Marmur, <u>J</u>. <u>Mol</u>. <u>Biol</u>., <u>33</u>, 123 (1968).

[37] N. B. Hecht, M. Bleyman, and C. R. Woese, <u>Proc</u>. <u>Nat</u>. <u>Acad</u>. <u>Sci</u>. <u>USA</u>, <u>59</u>, 1278 (1968).

[38] M. Bleyman, M. Kondo, N. Hecht, and C. Woese, <u>J</u>. <u>Bacteriol</u>., <u>99</u>, 535 (1969).

[39] D. D. Brown and C. S. Weber, <u>J</u>. <u>Mol</u>. <u>Biol</u>., <u>34</u>, 661 (1968).

[40] K. D. Tartof and R. P. Perry, J. Mol. Biol., 51, 171 (1970).

[41] D. M. W. Kritikar and A. Kaji, J. Biol. Chem., 243, 5345 (1968).

[42] J. D. Mandell and A. D. Hershey, Anal. Biochem., 1, 66 (1960).

[43] D. Gillespie and S. Spiegelman, J. Mol. Biol., 12, 829 (1965).

[44] H. A. Raué and M. Gruber, FEBS Letters 8, 45 (1970).

[45] R. Röschenthaler, M. A. Devynck, P. Fromageot, and E. J. Simon, Biochim. Biophys. Acta, 182, 481 (1969).

[46] H. I. Robins and I. D. Raacke, Biochem. Biophys. Res. Commun., 33, 240 (1968).

[47] R. Loeser, R. Roschenthaler, and P. Herrlich, Biochemistry, 9, 2364 (1970.

[48] Y. Moriyama, P. Ip, and H. Busch, Biochim. Biophys. Acta, 209, 161 (1970).

[49] G. G. Brownlee and F. Sanger, J. Mol. Biol., 23, 337 (1967).

[50] M. Cannon and E. G. Richards, Biochem. J., 103, 23c (1967).

[51] A. C. Peacock and C. H. Dingman, Biochemistry, 6, 1818 (1967).

[52] E. G. Richards, J. A. Coll, and W. B. Gratzer, Anal. Biochem., 12, 452 (1965).

[53] U. E. Loening, Biochem. J., 102, 251 (1967).

[54] L. Ornstein, Ann. N. Y. Acad. Sci., 121, 321 (1964).

[55] B. J. Davis, Ann. N. Y. Acad. Sci., 181, 404 (1964).

[56] J. M. Adams, P. G. N. Jeppesen, F. Sanger, and B. G. Barrell, Nature, 223, 1009 (1969).

[57] B. R. Jordan, J. Mol. Biol., 55, 423 (1971).

The manuscript was completed in November 1970. No fundamentally new methods have been reported since. A refinement of the Sephadex method was described by the laboratories of M. Nomura and V. A. Erdmann [Proc. Nat. Acad. Sci. USA 68, 2932 (1971) and Molec. Gen. Genetics 114, 89 (1971)].

Chapter 10

SYNCHRONIZATION OF MAMMALIAN CELLS BY SIZE SEPARATION

Charles A. Pasternak
Department of Biochemistry
Oxford University
Oxford, England

I.   INTRODUCTION

Synchronization of exponentially growing cells may be
divided into methods that (i) induce synchrony by bringing all
cells to a particular stage of their life cycle, and methods

that (ii) <u>select</u> those cells that happen to be at a particular

stage. Nongrowing cells are generally already partly synchron-

ized in that they are between M* and S phases, i.e., in G_0 or

G_1 [1, 2] (The abbreviations used for the phases of the cell

cycle are those in common use, e.g., see [1].).

   (i) <u>Induction</u> methods [3-5] include (a) the use of specific

chemicals [6, 7], such as inhibitors of DNA synthesis that arrest

cells in S phase but allow other cells to proceed around the

cell cycle to the G_1/S boundary, and (b) the effect of a sudden

temperature change [8] like chilling, which is likewise thought

to affect cells in S phase more than those at other stages.

Subsequent removal of the inhibitor or restoration of normal

temperature results in partial synchrony, which may be improved

by repetition of the whole process. The advantage of induction is

that it is applicable to cells irrespective of whether they

grow in suspension or in monolayer, and that all the available

cells become synchronized; the disadvantage is that interference

with metabolic events may produce untoward effects, such as a

period of unbalanced growth [7, 9-11] or an alteration in the

length of certain phases of the cell cycle [4].

   (ii) <u>Selection</u> methods [3-5] have the advantages that

cells are not subject to adverse conditions and that the cells

that are obtained may be analyzed immediately, instead of hav-

ing to progress through one cell cycle, with the inevitable

decrease in the sharpness of synchrony. Methods include

(a) mitotic harvesting, (b) selective killing, and (c) selection
by size. (a) Mitotic harvesting [12, 13], which is applicable
only to cells that attach to the surface of the vessel in which
they are growing, depends on the fact that during mitosis cells
"round up" and can be removed from the remaining nonmitotic
cells simply by gentle shaking. The yield is naturally low,
but the mildness of the method makes it a useful one. By exposure of cells to an inhibitor of mitosis such as colchicine
prior to selection, the yield is greatly improved; however, the
disadvantages associated with chemical induction (see above)
then apply. The converse of selective harvesting is (b) selective killing. This method involves the use of lethal substances,
such as high concentrations of hydroxyurea [14] or high specific
activity [^{3}H]thymidine [15], which specifically kill cells in
S phase. The remaining cells are then partially synchronized,
but the great spread, as well as the presence of dead cells,
detracts from the method. (c) Selection by size [16, 17] depends
on the fact that just before division cells are twice as large
as immediately after division. A method that separates according to size, such as filtration or gradient centrifugation, may
therefore be used. Cells should be recently cloned, in order to
avoid a spread of sizes unrelated to the cell cycle. The advantage of the method is that most of the cells can be recovered
intact, and it is especially useful for cells that do not attach
to glass or plastic, cells for which mitotic harvesting is in-
applicable.

This chapter describes the use of two techniques of rate (velocity) centrifugation in order to separate cells according to size. Centrifugation in tubes (II, B) has the advantage of being easily carried out aseptically and is, therefore, useful if synchronously _growing_ cells are desired. Centrifugation in a zonal rotor (II,C) has the advantage that a greater quantity of cells can readily be processed, and that wall and other effects inherent in tube centrifugation are avoided [18].

The following methods have been worked out [19-21] for P815Y neoplastic mast cells [22] and were also found to work with HC [23] and BHK cells. A linear gradient of Ficoll is used, since high concentrations of sucrose [16, 17] are rather toxic to these and other cells [5]. Synchronization by tube centrifugation has been confirmed with P388 [24], P815X-2 [25], L [26], and LS cells [27]. The last two cell lines normally grow in monolayers; they were adapted to growth in suspension culture before the synchronization experiments. The method is adaptable in principle to any cell line, provided (i) that cell volume increases continuously throughout the cell cycle and (ii) that no fluctuations in density offset the change in volume.

In Section III are described some of the criteria that may be used to establish that the basis of separation according to size is indeed separation according to position in the cell cycle. Separated cells may then be studied by various techniques, such as chemical [21, 28], enzymic [29], or immunological [30]

analysis. Exposure to radioactive tracers before separation
is useful to ensure that all the cells under study are in an
identical milieu.

II.   SEPARATION OF CELLS

A.   Preparation of Cells

P815Y cells are grown statically or with stirring at 37°C
in Medium ("Medium" is Fischer's [31] medium (Grand Island
Biological Co., New York) without serum; "growth medium" has
serum added) containing 10% horse serum, from about 1×10^5 to
1×10^6 cells/ml. One-liter screw-capped Erlenmeyer flasks
containing 250 ml of growth medium (static) or 2- to 5-liter
spinner flasks [32] filled with growth medium (stirred) are
used. Cells are harvested by centrifugation (5 min at 500 x g)
and washed in Medium. Exponentially grown P815Y cells have a
mean cell volume of 1200 μm^3 [21] and a doubling time of 15 hr,
of which 0.9 hr is spent in M, 4.5 hr in G_1, 6.5 hr in S, and 3 hr
in G_2 [33].

For analysis of DNA synthesis, cells are most conveniently
pulsed with [^{3}H]- or [^{14}C]thymidine before separation. A suspension of cells ($1-3 \times 10^7$ cells/ml) in warm medium is exposed
to 1 μCi/ml of thymidine (e.g., (Me-^{3}H]thymidine; 5 Ci/mmole)
for 30 min. The suspension is centrifuged and unincorporated
isotope is removed by washing. The speed during centrifugation
should not exceed 500 x g. Other synthetic activities during

the cell cycle may be measured in a similar manner with appropriate isotopes [21].

When growth of separated cells is to be followed (see III, A),
all operations are carried out aseptically.

B. Tube Centrifugation

A swing-out (swinging-bucket) rotor (No. 62301) in an MSE
Mistral 6L centrifuge, or its equivalent, is used. A linear
gradient (40 ml) of 5 to 10% Ficoll in Medium is prepared [34] in
a plastic centrifuge tube. Cells ($0.5-2 \times 10^8$ in 1 ml of medium)
are layered on top and stirred lightly. The tube is centrifuged
at 80 x g for 3-7 min; the length of time is inversely proportional to the cell load. The bottom of the tube is punctured
and 1-ml fractions are collected. For subsequent growth of
separated cells, the temperature is kept at or above that of
the room, and the gradient mixer and centrifuge tube are autoclaved before use. Otherwise the temperature is kept at 5°C;
Medium may then be replaced by 0.9% NaCl buffered with 50 mM
Tris·HCl, pH 7.4.

C. Zonal Centrifugation

An MSE type-A zonal rotor in an MSE Mistral 6L centrifuge,
or its equivalent, is used. A linear gradient of 1 liter of
2-10% Ficoll (a dextran polysaccharide) in 0.9% NaCl, buffered
at pH 7.4 with 50 mM Tris·HCl, is generated with an MSE fixed-
profile gradient former. The rotor is accelerated to 300 rpm,
and the gradient is pumped into the rotor from the edge, at

50 ml/min, followed by an underlay of 15% Ficoll; since cells do not come into contact with this underlay, 15% sucrose may be used instead. Cells (0.5-1 x 10^9 in 25 ml of 1% Ficoll in Medium) are pumped into the center of the rotor, followed by an overlay of 80 ml of 0.9% NaCl - 50 mM Tris·HCl (pH 7.4). The cell band, which is clearly visible, should be about 8 cm from the rotor center. The rotor is accelerated to 500 rpm for 12 to 15 min, depending on the cell load; the main cell band should have reached 10-11 cm from the rotor center. The rotor speed is decreased to 300 rpm, and fractions (about 25 ml) are collected from the center by pumping in more underlay solution to the edge of the rotor at 70 ml/min. All operations are done at 4°C.

III. CRITERIA OF SYNCHRONY

A. Cell Division

Cell fractions are diluted 5-fold in warm Medium and washed free of Ficoll. The cell density is further reduced to about 1 x 10^5 cells per ml of growth medium, and suspensions are incubated at 37°C in screw-capped tubes or flasks. A Coulter counter is used to estimate cell numbers. Cells near the top of the gradient should exhibit a lag before division, whereas cells near the bottom should divide without lag (Fig. 1). Mitotic index [35] may also be measured; it should be greater than 5-fold higher during mitosis, when cell number is increasing, than during interphase.

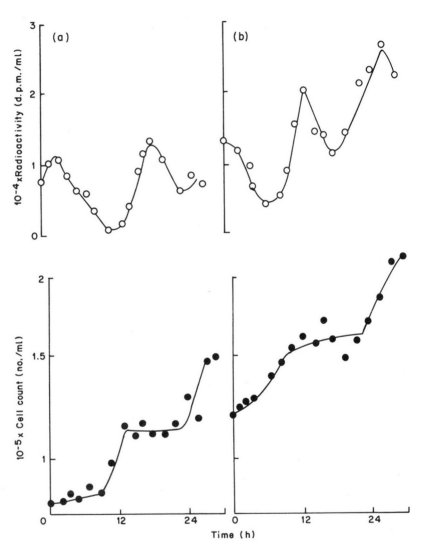

Fig. 1. Growth and [³H]thymidine incorporation of synchronous cells. Cells (1.5 x 10⁸) were separated by tube centrifugation, and pooled fractions from the upper region (18.7-19.0 cm
from rotor center) and lower region (22.0-22.3 cm from rotor center)

B. DNA Synthesis

(i) Prelabeled cells (II,A) are analyzed directly; a
convenient method is to add a sample (e.g., 0.1-1 ml) to ice-
cold 5% trichloroacetic acid (e.g., 1-10 ml), and to filter with
washing through Whatman fiberglass filters (GF/C, 2.1 cm diameter),
and to assay the radioactivity of dried filters by scintillation
counting (Figs. 2 and 3). Alternatively, the percentage of
cells that have incorporated [^{3}H]thymidine may be estimated by
autoradiography as follows: Cells are applied to microscope
slides, fixed in methanol, and dried in air. The slides are
dipped in warm Kodak Nuclear Track Emulsion NTB2 or its equivalent, dried in a stream of warm air, and kept at 4°C for one or
two weeks. Development in Kodak D-19b developer or its equivalent is followed by staining with Harris's hematoxylin (Fig. 2).

(ii) DNA synthesis during synchronous growth. Samples
(1 ml) containing 1-2 x 10^5 cells are removed at intervals and
exposed to 2 μCi of [^{3}H]thymidine for 30 min; incorporated
radioactivity is assayed by filtration in trichloroacetic acid
as above. Exposure to such pulses gives sharper changes (Fig. 1)
than does a continuous exposure to isotope [19]. Autoradiography (see above) of labeled cells may also be performed.

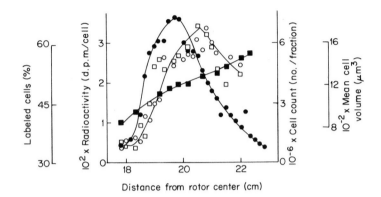

Fig. 2. Analysis of prelabeled cells by tube centrifugation. Cells (1 x 10^8 in 10 ml) exposed to 10 μCi of [^{3}H]thymi-
dine for 30 min were separated and analyzed as described in
Section III. ●, cell number; ■, mean cell volume; 0, ^{3}H-labeled
DNA; □, % of cells labeled with ^{3}H, determined autoradiographi-
cally.

C. DNA Content

Another criterion of successful separation is the DNA
content per cell. Samples containing 1-2 x 10^6 cells are twice
washed with ice-cold 5% trichloroacetic acid and extracted into
2 ml of 5% trichloroacetic acid at 100°C for 15 min; released
deoxyribose is measured [36]. Unlike other macromolecular constituents, DNA content should double fairly sharply (Fig. 4),
and thus allow one to identify the cells that are in S phase.

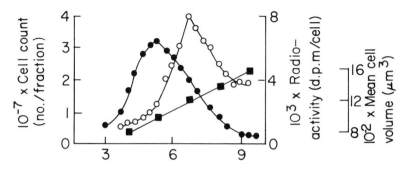

Fig. 3.   Analysis of prelabeled cells separated by zonal centrifugation.   Cells (1×10^9 in 25 ml) exposed to 20 µCi of [^{14}C]thymidine for 30 min were separated and analyzed as described in Section III. ●, cell number; ■, mean cell volume; O, ^{14}C-labeled DNA.

D. Cell Size

The size distribution of separated cells is obtained by counting cells, suitably diluted in 0.9% NaCl - 50 mM Tris·HCl (pH 7.4), at different threshold values with a Coulter counter (model A), with a 100-µm orifice.   Alternatively, the Volume Converter Accessory attached to the model Z_B may be used.   Polystyrene-divinylbenzene latex beads (12 µm diameter), paper-mulberry pollen (13.5 µm diameter), and silver-birch pollen (22.9 µm diameter) are used to calibrate the counter.   The mean cell volume is calculated from the distribution curve by the group-deviation method [37]; a continuous increase in cell size (Figs. 2 and 3) along the gradient should be obtained.

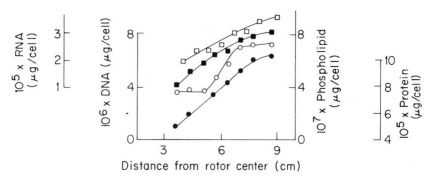

Fig. 4. Analysis of cells separated by zonal centrifugation. Cells (1 x 10^9 in 25 ml) were separated, and analyzed
as described in Section III. The content of total protein, ●,
RNA, □, and phospholipid, ■, is shown for comparison with that
of DNA, O.

ACKNOWLEDGMENT

I thank my colleague Dr. A. M. H. Warmsley for his skill
in pioneering these experiments.

REFERENCES

[1] R. Baserga, Cell Tissue Kinetics, 1, 167 (1968).

[2] R. A. Tobey and K. D. Ley, J. Cell Biol., 46, 151 (1970).

[3] E. Stubblefield, in Methods in Cell Physiology, Vol. 3
(D. M. Prescott, ed.), Academic Press, New York, 1968, p. 25.

[4] J. M. Mitchison, The Biology of the Cell Cycle, Cambridge
Univ. Press, London and New York, 1971, p. 25.

[5] A. H. W. Nias and M. Fox, Cell Tissue Kinetics, 4, 375
(1971).

[6] D. F. Petersen, R. A. Tobey, and E. C. Anderson, Fed. Proc.,
28, 1771 (1969).

[7] G. C. Mueller, Fed. Proc., 28, 1780 (1969).

[8] P. N. Rao and J. Engleberg, in Cell Synchrony, (I. L.
Cameron and G. M. Padilla, ed.), Academic Press, New York, 1966,
p. 332.

[9] J. H. Kim, S. H. Kim, and M. L. Eidinoff, Biochem. Pharmacol.,
14, 1821 (1965).

[10] G. P. Studzinski and W. C. Lambert, J. Cell Physiol., 73,
109 (1969).

[11] J. J. M. Bergeron, Biochem.J., 123, 385 (1971).

[12] T. Terasima and L. J. Tolmach, Exp. Cell Res., 30, 344 (1963).

[13] D. F. Petersen, E. C. Anderson, and R. A. Tobey, in Methods
in Cell Physiology (D. M. Prescott, ed.), Vol. 3, Academic Press,
New York, 1968, p. 347.

[14] W. K. Sinclair, Science, 150, 1729 (1965).

[15] G. F. Whitmore and S. Gulyas, Science, 151, 691 (1966).

[16] J. M. Mitchison and W. S. Vincent, Nature, 205, 987 (1965).

[17] R. Sinclair and D. H. L. Bishop, Nature, 205, 1272 (1965).

[18] N. G. Anderson, Science, 154, 103 (1966).

[19] J. J. M. Bergerson, A. M. H. Warmsley, and C. A. Pasternak,
FEBS Lett., 4, 161 (1969).

[20] A. M. H. Warmsley, J. J. M. Bergeron, and C. A. Pasternak,
Biochem. J., 114, 64P (1969).

[21] A. M. H. Warmsley and C. A. Pasternak, Biochem. J., 119,
493 (1970).

[22] T. B. Dunn and M. Potter, J. Nat. Cancer Inst., 18, 587 (1957).

[23] J. Furth, P. Hagen, and E. I. Hirsch, Proc. Soc. Exp. Biol. N.Y., 95, 824 (1957).

[24] S. R. Ayad, M. Fox, and D. Winstanley, Biochem. Biophys. Res. Commun., 37, 551 (1969).

[25] R. Schindler, L. Ramseier, J. C. Schaer, and A. Grieder, Exp. Cell Res., 59, 90 (1970).

[26] H. R. MacDonald and R. G. Miller, Biophys. J., 10, 834 (1970).

[27] S. Shall and A. J. McClelland, Nature New Biology, 229, 59 (1970).

[28] M. E. Cross, Biochem. J., In press.

[29] A. M. H. Warmsley, B. Phillips, and C. A. Pasternak, Biochem. J., 120, 683 (1970).

[30] C. A. Pasternak, A. M. H. Warmsley, and D. B. Thomas, J. Cell Biol., 50, 562 (1971).

[31] G. A. Fischer and A. C. Sartorelli, Meth. Med. Res., 10, 247 (1964).

[32] J. Paul, in Cell and Tissue Culture, 3rd ed., E. & S. Livingstone, Edinburgh, 1965, p. 262.

[33] J. J. M. Bergeron, A. M. H. Warmsley, and C. A. Pasternak, Biochem. J., 119, 489 (1970).

[34] T. Salo and D. M. Kouns, Analyt. Biochem., 13, 74 (1965).

[35] R. A. Tobey, D. F. Petersen, E. C. Anderson, and T. T. Puck, Biophys. J., 6, 567 (1966).

[36]   K. Burton, <u>Biochem. J.</u>, 62, 315 (1956).

[37]   J. Stanley, <u>The Essence of Biometry</u>, McGill Univ. Press, Montreal, 1963, p.23.

AUTHOR INDEX

Numbers in brackets are reference numbers and indicate that an author's work is referred to although his name is not cited in the text. Underlined numbers give the page on which the complete reference is listed.

A

Aaronson, A. I., 169 [8], 183 [8], 184 [8], <u>185</u>

Aaronson, S. A., 130 [6], 130 [9b], 130 [10b], 130 [12], 131 [10b], 131 [12], 134 [10b], <u>142</u>

Aaslestad, H. G., 20 [10], 44 [10], 58 [10], <u>60</u>

Abrell, J. W., 139 [10a], <u>142</u>

Adams, J. M., 238 [56], <u>245</u>

Alberts, B. M., 147 [14], <u>156</u>

Amaldi, F., 219 [18], 221 [33], 229 [33], <u>242</u>, <u>243</u>

Anderson, E. C., 248 [6], 249 [13], 253 [35], <u>259</u>, <u>260</u>

Anderson, N. G., 250 [18], <u>259</u>

Anfinsen, C. B., 66 [2], <u>91</u>

Angles, C. J., 113 [25], <u>125</u>

Apirion, D., 169 [9], <u>185</u>

Artman, M., 184 [39], <u>186</u>

Atkinson, M. R., 157 [5], <u>165</u>

Attardi, G., 219 [18], 221 [33], 229 [33], <u>242</u>, <u>243</u>

Aubert, M., 219 [24], 220 [24], 220 [28], 228 [28], 229 [28], 231 [28], <u>242</u>, <u>243</u>

August, J. T., 51 [24], 52 [24], 55 [29], <u>61</u>, <u>62</u>

Aviv, H., 189 [5], 189 [8], 207 [5], 209 [22], 209 [24], 209 [27], <u>214-216</u>

Ayad, S. R., 250 [24], <u>260</u>

B

Baker, R. F., **174** [23], 175 [23], <u>186</u>

Baltimore, D., 58 [35], 58 [39], 58 [41], 59 [41], 62, 127 [1], 130 [1], <u>130</u> [2], 139 [2], <u>141</u>, 145 [5], 149 [5], <u>155</u>

Banerjee, A. K., 55 [29], <u>62</u>

Barrell, B. G., 49 [19], <u>61</u>, 219 [19], 219 [20], 238 [56], <u>242</u>, <u>245</u>

Baserga, R., 248 [1], <u>258</u>

Beaudreau, G. S., 131 [8], <u>142</u>

Bellett, A. J. D., 197 [14], <u>215</u>

Benson, R. H., 103 [15], 120 [15], <u>124</u>

Berg, P., 98 [7], <u>124</u>

Bergeron, J. J. M., 248 [11], 250 [2], 251 [33], <u>259</u>, <u>260</u>

Bergerson, J. J. M., 250 [19], 255 [19], <u>259</u>

Bergquist, P. L., 67 [4], 69, 70 [4], <u>91</u>

Berkower, I., 145 [3], 146 [9], 154 [20], <u>155</u>, <u>156</u>, 158 [10], 159 [10], 162 [10], <u>165</u>

263

SUBJECT INDEX

A

Acetonitrile, 5
Actinomycin D, 76, 83, 219
Agarose column, 148
Alkaline phosphatase, 154
Alumina, 221
α-Amanitin, 110
Amino acid starvation, 170, 173, 184
Ampholytes, 138
Annealing, 42, 44, 55
Ascites cells, 189, 192, 197
Autoradiography, 255
Auxotrophs, 173
8-Azaguanine, 170

B

Barium chloride, 6
Bentonite, 67
Bis-acrylamide, 35
Blastocladiella emersonii, 218
Bovine serum albumin, 15, 24, 116, 121, 140, 146
Brij, 58

C

Calf-thymus DNA, 115
Cell division, 253
Cell size, 257
Cesium chloride, 54
Chicken embryo, 134
Chloramphenicol, 168, 220
Chromatin, 102
Chromatography, 25
Cobalt, 170
Conditional lethal mutants, 175

Corn seedlings, 99
Coulter counter, 257
Cysteine, 174

D

DEAE-cellulose, 29, 53, 108, 133, 148, 160, 233
DEAE-cell-lose-urea, 30
DEAE-cellulose-urea column chromatography, 46
DEAE-Sephadex, 235
Denaturation, 41
Density differences, 41
Deoxycholate, 223
Deoxyribonuclease, 74
Deproteinization, 65, 90
Detergents, 23
Diethylpyrocarbonate, 67
DNA cellulose column, 148
DNA-dependent DNA polymerase, 134
DNA ligase, 162
DNA polymerase I, 157
DNA polymerase II, 157
DNA polymerase II assay, 158
DNA polymerase II properties, 162
DNA polymerase II purification, 159
DNA-RNA hybrids, 127, 145
DNase, 71, 79, 158
DNA synthesis, 247, 255
Double-strandedness, 40, 57

E

EDTA, 24, 75, 81, 101
Electrofocusing, 137
Electropherogram, 84
EMC RNA, 203